STATISTICAL

ANALYSIS

IN GEOGRAPHY

PRENTICE-HALL INTERNATIONAL, INC., *London*
PRENTICE-HALL OF AUSTRALIA, PTY. LTD., *Sydney*
PRENTICE-HALL OF CANADA, LTD., *Toronto*
PRENTICE-HALL OF INDIA PRIVATE LTD., *New Delhi*
PRENTICE-HALL OF JAPAN, INC., *Tokyo*

STATISTICAL
ANALYSIS
IN GEOGRAPHY

Leslie J. King

Professor of Geography
The Ohio State University

PRENTICE-HALL, INC.,
ENGLEWOOD CLIFFS, N.J.

Current printing (last digit):

10 9 8 7 6 5 4 3 2 1

Library of Congress Catalog Card Number 69-15045
Printed in the United States of America

To JOBBY and MAC
my teachers

PREFACE

Over the past seven years the author has been concerned with the teaching of quantitative methods to beginning graduate students in geography. This book represents an attempt to summarize some of the major themes which have been emphasized in these teaching efforts and to review some of the substantive work which has been used to illustrate these themes. The author's experience is that the material included in the book can be presented conveniently over two ten-week quarters to beginning graduate students who have a minimum of training in statistics and mathematics.

The accelerated development of theoretical and quantitative work in geography over the past decade has been most conspicuous in the fields of human geography. One consequence of this development has been the revival of certain lines of methodological debate such as on the issue of whether geography should be concerned more with the particular and unique or with the general and universal. The materials reviewed in this book and the very subject matter of the book itself support only the second of these two positions. It is taken for granted that geography as a discipline has a strong social and behavioral science component and that the use of statistical techniques therefore is an appropriate research strategy to pursue in the testing and verification of theoretical constructs.

On a more detailed level, this book makes no serious attempt to outline any strong methodological position. Although it is apparent that most of the studies reviewed here are consistent with the theme of spatial analysis, there is no reason why the so-called "man-land" and "areal differentiation" studies should not be amenable to statistical analysis. Some of the recent work on environmental perception lends weight to this point.

In developing the content of this book the author has benefited from attendance at the N.S.F.-sponsored seminars in Computer Applications in Geography and in Spatial Statistics held at Northwestern University, and to the organizers and lecturers at these seminars the author owes a debt of gratitude. Also, in 1965 the author was fortunate in having the opportunity to participate with Prof. Edward J. Taaffe in organizing an N.S.F. Institute on Quantitative Methods in Geography. Those who have worked with Ned and sensed the enthusiasm and insight with which he approaches the problem of developing scientific research in the discipline will understand the sincere vote of thanks which is accorded him here.

Many other individuals have contributed to this book either directly or indirectly. Professors J. Rayner and E. Casetti at Ohio State, and Prof. P. Haggett and Dr. D. Harvey at Bristol read parts or all of the manuscript and offered many constructive criticisms. The influence of Prof. Edwin N. Thomas, who supervised much of the author's graduate work in quantitative research, is woven throughout the text, although only a few will recognize it and Ed probably would disclaim it. No one can attempt to write a book such as this without being conscious of the general debt owed to Prof. W. L. Garrison, who by way of rigorous scholarship demonstrated the utility of scientific research in human geography. A great number of the studies reported on in this volume were completed by scholars who studied under Garrison. Finally, the author would like to acknowledge the stimulation he has received from discussions on particular topics with Brian Berry, Leslie Curry, Michael Dacey, Peter Gould, Duane Marble, Gunnar Olsson, and Waldo Tobler. The errors which persist in the book remain the author's responsibility alone.

A vote of thanks is accorded Miss Carol Fritz who showed a great deal of patience and professional skill in typing the manuscript. My wife Doreen gave invaluable support by way of her encouragement and sacrifices.

CONTENTS

1

2

3

4

5

6

7

PRINCIPAL COMPONENTS AND FACTOR ANALYSIS IN GEOGRAPHIC RESEARCH, 165

8

CLASSIFICATION AND REGIONALIZATION PROBLEMS, 194

9

EMERGING TRENDS AND FUTURE PROSPECTS, 216

Appendices

A

B

STATISTICAL
ANALYSIS
IN GEOGRAPHY

INTRODUCTION

Over the past decade there has been an increasing
use of statistical analysis in geographic research.
This development has paralleled the emergence in
most social sciences of a greater emphasis upon
scientific method and the associated languages of
mathematics and statistics and, as in many of these
other disciplines also, it has met with considerable
opposition.

This opposition has been strongest in the non-
physical fields of geography and has focused upon
such themes as the "idiographic versus the nomo-
thetic" approach, the "qualitative nature of many
phenomena as opposed to the quantitative requirements of science," "aggregative versus behavioral
approaches," and the distinction between "correlation
and causation." Many of these arguments have considerable merit and are far from being resolved. The
discussion of such issues, however, is pursued more
appropriately in the context of either methodological

1

tracts or substantive research studies, and this text falls in neither of these categories. Rather it attempts a review of the applications of statistical analysis in geography to date, pointing up some of the achievements along these lines and underlining the relative strengths and weaknesses of different analyses.

A review of this type appears all the more desirable in light of the increasing evidence that many of the statistical techniques which have been used previously in geography are inappropriate for the particular problems posed in geographic research and that worthwhile achievements will be realized only when more formal and often original mathematical analysis has been completed. The bases for the claims made along these lines are not widely understood within the discipline, and it is hoped that this text will at least make a contribution in this respect—to bring into clearer focus for a larger audience some of the problems associated with statistical analysis in geographic research.

In discussing these problems, however, care must be taken that one does not lose sight of the whole forest! It is certainly the case that results obtained from statistical analyses in geography often have been weakened by their complete disregard of many technical problems, some of which are discussed in this text. At the same time, there have been significant net gains in the understanding of locational distributions and spatial processes as a consequence of statistical analyses, which now form part of the published literature of the discipline. Hopefully, therefore, the review which is attempted in the following chapters will give the reader some appreciation of the breadth and depth of statistical analysis in geography. For the most part, the examples are drawn from economic and urban geography, but occasional reference is made to studies in other areas of geographic and social science research.

At the outset, it is as well to view the trend toward a greater use of statistical analysis in geography against the background of broader developments in science. Whatever the discipline, the scientific method demands precise identification of problems, accurate description, and the formulation and development of hypotheses and models which might serve as explanations of reality. The pursuit of scientific inquiry has a cyclical nature, that is to say, the scientist proceeds from observed facts to the formulation of hypotheses or models, the logical consequences of which are then tested against the facts. The agreement or discrepancy between the "expected" facts or predictions and the observable facts is noted and the cycle of hypothesis formulation and so on commences again. Throughout the pursuit of this inquiry the language of mathematics enjoys a preeminent role largely by virtue of its powerful logical properties. In this book we are not concerned with pure mathematical analysis and precise deterministic relationships. We are interested in statistical analysis, that is to say, the analysis of situations in which there is variation present in the measured properties being dealt with and the relationships being specified.

Kemeny (1959) has noted how the notion of indeterminacy and the associated statistical laws have found expression at two levels within science. In the first place, it has become apparent in many areas of research that there are errors associated with measurement and calculations and that even if precise deterministic laws exist, predictions will be off by some small amount as a consequence of these errors. Recognition of this fact gave rise to the "theory of errors" and to the development of a formidable body of statistical techniques including estimation procedures and the method of least squares. Some of these topics are touched upon in this book; however, it is important to appreciate that questions of measurement and sampling error have not yet been fully investigated by social scientists. Physical scientists, including physical geographers who have been involved in the actual collection and recording of their data by measurement procedures, have been aware of these error problems for some time. The concern of the social scientist over similar problems is a more recent development and unfortunately it has yet to be emphasized in geography. Reliance is placed all too frequently upon census and other published data for which the error terms are not known.

On the second level, statements of indeterminacy occur in science, particularly it seems in social science, when phenomena under consideration appear to act or to be distributed in a random manner subject to no known laws. This question of the operation of chance factors, and the indeterministic approach to scientific problems which it implies, is a very complex philosophical problem. The reading of a book such as Kemeny (1959) is suggested as an excellent introduction to the subject.

It is possible in geography, as in any science, to proceed deductively and to derive deterministic models which allow for no variance in the relationships among the phenomena under study. The agricultural land-rent model can be cited as an example of a deterministic model. In this case,

$$R = Y(p) - E - Y(f)(k),$$

where R is the rent per unit of land area, p is the market price for a unit of production, Y is the yield in units of production per unit of land area, E is production costs per unit of land area, f is the transport rate as a cost per unit of production per unit distance, and k is the distance separating the land in question from the market. The model appears as a precise mathematical function and if the terms on the right are known, then R is determined.

If we were to collect some empirical data on the factors in question, however, we would find almost certainly that this precise relationship does not hold. As a result of this data analysis, we may be led to conclude that the relationship among the factors involves some error terms. In another context, for example, the relationship between the population size of towns (Y) and their functional complexity (X) is given usually as

$$Y = a + bX + e,$$

where a and b are empirically derived constants and e is the error term. Now the problem is to interpret this error term. It may be that the relationship between the factors in question is a fairly precise one and e is ascribed to factors which have not yet been considered. Alternatively, the error may reflect chance or random influences whose outcomes on any particular occasion can never be predicted.

The suggestion that chance might be important in scientific inquiry has prompted the development of a large body of statistical techniques and mathematical theories. Again it is possible to recognize various levels or stages by which this development has proceeded. Neyman (1960), for example, has suggested that there have been at least three important stages. The first was in the late nineteenth century when attention was focused upon the problem of describing in mathematical terms the variation which resulted from the operation of the chance mechanisms. In this endeavor, the work of K. Pearson is rated as outstanding and his system of theoretical frequency curves and methods of curve fitting still command attention. At different points in this text, reference is made to attempts by geographers to fit different mathematical curves to observed geographical data. The work on the urban rank-size problem, the location pattern of towns within an area, and the pattern of migration flows are three such examples.

A second phase in the emergence of the indeterministic approach and the development of statistical analysis centered around the work of R. A. Fisher during the first half of this century. In this period, much of the existing knowledge on experimental design and the testing of statistical hypotheses was formalized, although many important revisions and even alternative approaches have since been proposed. In this analysis, known generally as statistical inference, emphasis is placed upon the use of samples in testing hypotheses or drawing inferences concerning the larger parent groups or populations from which the samples were selected. The logical bases upon which the hypotheses then are either accepted or rejected are provided by the mathematics of probability theory.

This text considers a number of questions of statistical inference as they have appeared in the geography literature. The majority of these analyses in geography have emphasized the testing of hypotheses concerning the nature of location patterns, the extent to which different phenomena are interrelated and associated over the surface of the earth, the importance and validity of certain regional constructs, and the significance of selected factors in differentiating between regions.

This book does not claim to be an exhaustive survey of statistical analysis

in geography. Indeed, it deliberately retreats from any detailed consideration of those models which illustrate what Neyman has described as the
third phase in the development of the indeterministic approach in science,
that of dynamic indeterminism. This more recent, and certainly more
demanding in terms of the mathematics involved, phase emphasizes the study
of evolutionary processes subject to random change. It is a phase of development into which geography is now entering, and the "frontier" nature of
this geographic research is reflected in the work currently being undertaken
on population migration, cultural diffusion, drainage basin evolution, climatic
change, and urban systems, to name but a few widely separated areas. At
present, these geographic models are crudely fashioned but even in this form
they are suggestive of fruitful approaches for future geographic research.
A fuller exposition of these approaches undoubtedly will be forthcoming
in the future; in this text, they are given only brief attention in the final
chapter.

It is stressed that this is not intended primarily to be a statistical techniques
book. It contains no mathematical theory and only at certain points does
it refer to relevant aspects of the underlying theory. As noted earlier, it
is first and foremost a review of statistical applications in geography. This
review is undertaken in the hope that the geography student, by familiarizing
himself with the strengths and weaknesses of these applications, may become
more critical in his own use of the techniques and, in so doing, contribute
to a more rigorous pursuit of geographic inquiry.

Hopefully then, the use of this book in teaching will be supplemented
by reference to any one of the number of excellent introductory statistical
texts which are available. The author in his own teaching of quantitative
applications in geography has found that a book such as Hoel (1960) can
be handled by most first-year graduate students and that Fraser (1958)
and Johnson and Leone (1964) are within the grasp of the better students.
These books provide technical discussions of most of the methods covered
in the first six chapters of this text. For the multivariate analysis in the
later chapters there are no comparable references, although Kendall (1957)
and Morrison (1967) are excellent intermediate-level texts.

The division of the material into chapters reflects the major thrusts of
quantitative research in geography to date. The first two substantive chapters
merely review some standard introductory topics in statistics and probability
theory and give references to their applications in geographic research.
Chapter 4 provides an overview of the methods of statistical inference and
areal point sampling. It is apparent from the few examples which are cited
that inferential questions have not been emphasized in geographic research
and it is suggested that this reflects in part the geographer's typical concern
with total populations rather than with samples drawn from them. The
emphasis, however, appears more historical than philosophical.

In Chapter 5 some specific issues in spatial analysis are emphasized, particularly the nature of random point processes. As noted in the text, the analytical results in this context are far more impressive than the empirical research findings. The following chapter on regression and correlation analysis is the longest in the book and provides the linkup with multivariate analysis. The length is justified if only in terms of the references which can be cited, for it is this line of analysis which has dominated statistical work in geography. Unfortunately, it is in this type of analysis that many technical problems related to spatial situations have become apparent and a section of this chapter is given over to a discussion of some of these problems.

These first six chapters of the book should be handled easily by most senior undergraduate majors and introductory graduate students.

The later sections of Chapter 6 and the last three chapters will require some familiarity with matrix algebra. An introduction to this subject is given in Appendix A. Obviously, no short treatment of multivariate analysis such as presented here will appear satisfactory. The aim in this book is simply to expose the reader to some of the variety of applications, while stressing the singularly geographic nature of the research problems. The work on regionalization which is discussed in Chapter 8 clearly illustrates this last point.

In the final chapter some of the contemporary trends in quantitative geographic research are mentioned. The growing emphasis on probabilistic formulations and the heightened concern for theory and deductive work which underlies this emphasis hold promise for much more fruitful quantitative work in geography in the future.

Throughout the text, an attempt has been made to identify readings and articles which best illustrate the geographer's use of the particular techniques. Additional readings can be drawn from the list of references which is given at the back of the book. Many of the articles discussed in this text are now more widely available as a result of the recent publication of an excellent book of readings by Berry and Marble (1967).

SUGGESTED READINGS

Burton, I. (1963), "The Quantitative Revolution and Theoretical Geography." *The Canadian Geographer*, Vol. 7, pp. 151–162.

Chorley, R. J. (1964), "Geography and Analogue Theory." *Annals*, Association of American Geographers, Vol. 54, pp. 127–137.

Haggett, P. (1965), *Locational Analysis in Human Geography*. New York: St. Martin's Press, Inc., pp. 1–27.

Kemeny, J. G. (1959), *A Philosopher Looks at Science*. Princeton: D. Van Nostrand Co., Inc., Chapters 1–10.

National Academy of Sciences—National Research Council (1965), *The Science of Geography*. Report of the Ad Hoc Committee on Geography, Earth Sciences Division.

Neyman, J. (1960), "Indeterminism in Science and New Demands on Statisticians." *Journal of American Statistical Association*, Vol. 55, pp. 625–639.

NUMERICAL DATA IN
GEOGRAPHIC RESEARCH

One of the issues raised in most discussions of quantitative analysis in geographic research is whether it is actually possible for the geographer to measure the properties or factors in which he is interested. The criticism often implied in such comments is not a very strong one but it is well to begin with a brief review of the ways in which measurement is achieved and how different forms of measurement have been utilized in geographic research.

2.1 MEASUREMENT IN GEOGRAPHY

The geographer's measurements typically relate to areas or points on the earth's surface or to aspects of the interaction between these areas or points. In all other respects, however, these measurements are no different than those made by scientists in other disciplines.

2

TABLE 2.1. BARBADOS FARMING CHARACTERISTICS

Attributes are grouped into *Crops* (1–42) and *Livestock* (43–48). Values are 0/1.

Farms	1 Sugar cane	2 Cabbage	3 Carrot	4 Cucumber	5 Cowpea	6 Pigeon pea	7 Beans	8 Egg plant	9 Okra	10 Pumpkin	⋯	21 Yams	22 Sweet potato	23 Maize	24 Cassava	25 Eddoes	26 Groundnuts	27 Ginger	⋯	41 Cherry	42 Akee	⋯	43 Cattle	44 Sheep	45 Goats	46 Pigs	47 Fowls	48 Equines
1	1	0	0	0	0	0	0	0	0	0	⋯	1	0	1	1	1	0	0	⋯	0	0	⋯	1	1	0	0	1	0
2	0	0	0	1	0	1	0	0	1	1	⋯	1	1	1	0	1	0	0	⋯	0	0	⋯	0	1	0	1	1	0
3	1	0	0	0	0	0	0	0	0	0	⋯	0	0	0	0	0	0	0	⋯	0	0	⋯	1	1	1	1	1	0
4	0	0	0	0	0	0	1	0	0	0	⋯	0	1	1	0	0	0	0	⋯	0	0	⋯	0	1	0	0	1	0
5	1	0	0	1	0	0	1	0	0	0	⋯	0	1	1	1	1	0	0	⋯	0	0	⋯	1	1	1	1	1	1
6	1	1	1	0	0	0	1	0	0	1	⋯	1	1	0	1	1	0	0	⋯	0	0	⋯	1	0	0	0	1	0
7	1	0	0	1	0	0	1	0	0	0	⋯	1	1	1	1	1	0	0	⋯	0	0	⋯	1	0	1	0	1	0
8	1	1	1	0	0	1	1	0	1	1	⋯	0	0	0	1	0	0	0	⋯	0	1	⋯	0	0	0	0	1	0
9	1	0	0	1	0	1	0	0	0	0	⋯	1	1	1	0	0	0	0	⋯	0	0	⋯	0	1	0	1	1	1
10	1	0	0	0	0	1	0	0	0	0	⋯	0	0	0	0	0	0	0	⋯	0	0	⋯	1	0	0	1	1	1
·	·	·	·	·	·	·	·	·	·	·	⋯	·	·	·	·	·	·	·	⋯	·	·	⋯	·	·	·	·	·	·
150	1	0	0	1	0	1	0	0	0	0	⋯	1	0	1	0	0	0	0	⋯	0	0	⋯	1	1	1	1	1	0

Source: Henshall and King (1966, p. 77).

At the simplest level, it is possible to introduce numbers into an analysis by way of a classification. A grouping of towns, farms, slopes, or persons, for example, into classes to which identifying numbers are assigned provides an effective quantification of the property or properties upon which the classification is based. This is *nominal* measurement and in its simplest form of only two classes, which are identified by *zero* and *one*, it has been used frequently in geography. In Table 2.1, for example, there is part of a data set relating to crop-farming characteristics in Barbados taken from a study by Henshall and King (1966). The observations in this case are the presence or absence of certain crops on a sample of peasant farms, the *ones* indicating that the particular crops are grown, the *zeros* that they are not.

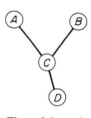

Similarly, some characteristics of the transportation network shown in Figure 2.1 may be expressed numerically. The existence of a direct two-way link between any two nodes on the network is shown as a *one* in Table 2.2, the absence of a connecting link as a *zero*. A node is not connected to itself, hence the zeros along the diagonal *AA–DD*. A rectangular array of numbers, as in Table 2.2, is referred to henceforth as a *matrix*. In transportation geography, the body of Table 2.2 is known as the "connectivity matrix." Kansky (1963) provides a discussion of many of these measures related to transportation networks.

Figure 2.1. A simple transport network.

It is worth noting here that a matrix is a special type of number which can be manipulated to yield additional results. The basic concepts of *matrix algebra* are outlined in Appendix A and a familiarity with these concepts will be assumed throughout the rest of this book. At some convenient time, therefore, the student should work through this introduction to the subject and consult some of the supplementary references given in that appendix.

In the area of transportation geography, it is often of interest to raise the connectivity matrix to some power. The square of the matrix, for example, gives the number of two-link connections between the pairs of nodes.

In nominal measurement it is assumed the classes are *mutually exclusive* with each observation belonging to one and only one class and also *exhaustive* in the sense that the full range of observations is catered to by the

TABLE 2.2. CONNECTIVITY MATRIX

Places	*A*	*B*	*C*	*D*
A	0	0	1	0
B	0	0	1	0
C	1	1	0	1
D	0	0	1	0

classification. The standard arithmetic operations of addition or multiplication have no meaning when applied to nominal measurements. It is possible, however, to derive other numerical values from the same data, such as simple counts of the occurrences of ones or zeros, which then can be used in subsequent analysis.

In nominal measurement there are no relationships among the observations other than the fact that they belong to different classes. Frequently something more is known about the observations and a ranking or ordering of them on a scale of size or preference is possible. The relationship of one observation to another may be expressed simply by qualitative statements such as "*A* seems higher than *B*" or "I prefer to shop at this center rather than at that one." Nevertheless, it is assumed that these statements allow for a ranking of the observations in either ascending or descending order. To each rank a number is assigned and these values are the numerical data.

In Table 2.3 there are two sets of rankings from an urban geography study of the Indian Punjab (Mayfield, 1963). On the occupational scale,

TABLE 2.3. INDIAN URBAN RANKINGS

	Occupation by gross Income level		Caste or social rank	
Low	1. Farm laborer	High	1. Brahmin and Jat Sikh	
	2. Services	↑	2. Sikh	
	3. Noncultivating production		3. Kashtriya	
	4. Transport		4. Vaisha	
	5. Tenant cultivator		5. Ramdasia Sikh	
	6. Commerce	Low	6. Shudra	
	7. Owner cultivator			
High	8. Landlord			

Source: Mayfield (1963, p. 44).

it should be noted, the number *one* identifies the lowest position, whereas on the caste scale it denotes the highest ranking. This poses no problems here, although if a set of individuals or areas were to be compared on several rankings, the comparison would be facilitated by having the scales numbered in the same direction.

The example in Table 2.3 is useful for making several points about ranking or *ordinal* scales. First, it is always assumed that the relationships among classes or observations are *asymmetric;* that is, if *A* is greater than *B*, then *B* cannot be greater than *A*. Secondly, the relationships are also *transitive* in the sense that if *A* is greater than *B* and *B* is greater than *C*, then *A* must be greater than *C*.

Again, it is important to note that the standard arithmetic operations are not defined for the numbers used on ordinal scales. In Table 2.3 the number 4 associated with the caste "Vaisha" does not mean that "Vaisha" is twice as high on the scale as "Sikh." Nor is it correct to assume that "Sikh" plus "Vaisha" is equal to "Shudra." The numbers simply denote order and the only meaningful relationship existing here between the numbers 4 and 2 is that the class denoted by 4 is understood to lie at a lower position on the caste scale than the class identified by 2.

Two other comments are made with respect to this example of ordinal data. First, the occupational scale is established on the basis of "gross income level," which itself is a different type of numerical data from those discussed so far. But it is not uncommon for rankings to be made on the basis of different types of measurements, indeed this approach may be preferred when the accuracy of the original measurements is questionable. As long as the limited properties of the numbers employed as rankings are appreciated and not confused with those of the original data, useful results can be obtained. Second, the rankings in Table 2.3 are derived with respect to a single criterion in each case. Multiple-item scaling has not been employed extensively in geography to date, but it undoubtedly will increase with the growing emphasis in some areas of the discipline on behavioral and perception factors which cannot be measured adequately on single-item scales. This point is reiterated in the final section of this book. El-Kammash (1963) has suggested the use of multiple-item scales in the study of economic development.

Apart from nominal and ordinal scales, measurement is made most commonly with respect to either an *interval* or *ratio* scale. It is this type of measurement which is basic to most of the standard arithmetic operations. On an interval scale there is a known distance separating any two numbers and this interval has meaning with respect to the property being measured. Intervals can be added and subtracted. The fact that the zero point is arbitrary, however, as in the case of the centigrade or Fahrenheit scales for temperature, means that the ratios of different values on the scale have no meaning. On a ratio scale, by contrast, the zero point is fixed or "true" and it is possible to compare not only the intervals but also the actual values or quantities on the scale. This is the case with the measurement of weight, distance, elevation, income, employment, and production, which are phenomena of interest not only to the geographer but to scientists in other fields. Hodge (1963) has discussed many of these questions of measurement in the context of urban planning research.

One final point noted here with respect to measurement is that measurement presupposes agreement as to what it is that is being measured! It is essential that good operational definitions of the concepts involved in any study be available before any measurement of the concepts is attempted. The fact that many of the concepts used in geography remain poorly defined at best has prompted numerous attempts at defining and measuring them by way of coefficients or ratios. The "location quotient," the "index of connectivity," and the "primacy ratio" are but a few of the examples which might be cited. What should be required of any ratio is that it appear a meaningful one, given the nature of the problem and the extent of existing knowledge and theory. Too often, complex ratios are used which suggest poorly defined concepts and which accomplish little more than might have been achieved by the use of simpler and more direct measurements.

2.2 REPRESENTATION OF GEOGRAPHIC DATA

The use of some simple notational and graphical schemes for representing sets of measurements is now reviewed. In this context, the term "set" is defined loosely as a collection of numbers, but this usage is not inconsistent with the more formal use of the term in mathematics. We shall be far less precise than the mathematician, however, in specifying the properties or characteristics of the sets.

In Table 2.4 there are data pertaining to the population sizes of cities in Canada. These data are presented in the form of a frequency distribution in Table 2.5. This is a simple tabular summary of the original data. The 157 observations have been grouped into a set of 11 population-size classes. For any one of these classes, designated by the symbol j, there is a frequency or count (f_j) of the number of cities contained in it. Instead of dealing with all 157 observations, it is necessary now to deal only with 11 representative points for the classes. These representative points usually are taken to be the class midpoint values ($\hat{X}_j$'s). This tabulation of the data involves some loss of information. For the ungrouped data in Table 2.4 the sum of the population size values (X_i) is given exactly as

$$\sum_{i=1}^{157} X_i,$$

where the uppercase Greek letter sigma denotes summation and the subscript i for the observation is used in specifying the range over which the summation is to apply. The use of this summation operator is explained in Appendix B. In this example, all 157 observations are to be summed. Hence

$$\sum_{i=1}^{157} X_i = 3860.$$

In the situation represented in Table 2.5, however, this sum can only be approximated since the magnitudes of the individual values are concealed. Assuming that the class midpoint values truly are representative of the observations contained in the different classes, then the sum will be approximated closely by the expression

$$\sum_{j=1}^{11} f_j \hat{X}_j.$$

TABLE 2.4. 1961 POPULATIONS FOR CANADIAN CITIES OVER 10,000 IN SIZE. CENTRAL CITIES OF METROPOLITAN AREAS EXCLUDED.

Figures in thousands				
112	33	22	16	12
96	32	22	15	12
84	32	22	15	12
78	31	22	15	12
67	31	22	15	12
62	31	21	15	12
57	31	21	14	12
56	31	21	14	12
55	30	20	14	11
54	30	20	14	11
53	30	20	14	11
51	29	20	14	11
50	29	20	14	11
48	28	20	14	11
47	28	19	14	11
47	28	19	14	11
47	28	19	13	11
45	27	19	13	11
45	27	19	13	11
44	27	18	13	11
44	25	18	13	11
43	25	18	13	11
41	25	18	13	11
40	25	18	13	11
39	24	18	13	11
38	24	18	13	10
36	24	17	13	10
35	24	17	12	10
34	24	17	12	10
34	24	16	12	10
34	23	16	12	10
			12	10

Source: 1961 Census of Canada, General Review Bull. 7.1–2, Table 3.

In the present example this sum is 3996.5. The discrepancy between this value and the corresponding one for the ungrouped data illustrates the loss of information associated with a grouping of data.

As in any classification, the class intervals for a frequency distribution should be mutually exclusive and exhaustive of the original set of observa-

TABLE 2.5.   FREQUENCY DISTRIBUTION OF CANADIAN URBAN DATA

Population size class (in thousands)	$\hat{X}_j$	f_j	p_j	$100p_j$	Cumulative $100p_j$
10–19	14.5	81	0.516	51.6	51.6
20–29	24.5	34	0.217	21.7	73.3
30–39	34.5	18	0.115	11.5	84.8
40–49	44.5	11	0.070	7.0	91.8
50–59	55.5	7	0.045	4.5	96.3
60–69	65.5	2	0.013	1.3	97.6
70–79	75.5	1	0.006	0.6	98.2
80–89	85.5	1	0.006	0.6	98.8
90–99	95.5	1	0.006	0.6	99.4
100–109	105.5	0	—	—	99.4
110–119	115.5	1	0.006	0.6	100.0
		N = 157	1.000	100.0	

Source: As for Table 2.4.

tions. The number of class intervals used is generally a subjective judgment on the part of the researcher, although there are some guidelines available. Huntsberger (1961, p. 10), for example, gives the following formula for estimating k, the number of class intervals to be used:

$$k = 1 + 3.3 \log n,$$

where n is the total number of observations and log represents the logarithm to the base 10.

A graphical representation of the frequency distribution above is given in Figure 2.2. A simple graph of this form, with the class intervals or midpoints measured on the horizontal axis (the abscissa) and the frequencies represented as columns measured on the vertical axis (the ordinate), is a *histogram*. A joining together of the class midpoint values at the top of the columns yields a *frequency polygon*.

The frequency distribution plotted in Figure 2.2 obviously is not symmetrical. In this case, where there is a "tail" of the distribution to the right, it is said to be *positively skewed*. If the tail is to the left, it is *negatively skewed*. For the most part, the frequency distributions encountered in geographic research show varying tendencies toward positive skewness, a fact which is significant for some of the inferential procedures to be introduced later. These demand particular assumptions of symmetry about the frequency distributions involved. As we shall see, there are a number of ways by which this problem can be circumvented.

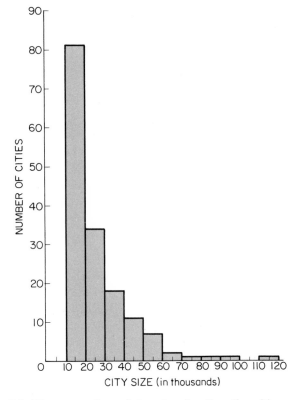

Figure 2.2. Histogram of population sizes for Canadian cities over ten thousand, 1961.

Given the frequency distribution in Table 2.4, it is a simple matter to express each frequency (f_j) either as a proportion (p_j) or as a percentage $(100p_j)$ of the total number of observations N (Table 2.5).

A cumulative summing of the f_j, p_j, or $100p_j$ values gives the number, proportion, or percentages of observations which fall either above or below specified values. In Table 2.5, for example, 84.8 percent of the 157 cities have a population equal to or less than 39,000.

The cumulative percentage distribution can be calculated for ungrouped data which have been ranked by value and for grouped data as well. In either case, it is possible to speak of an observation's position with regard to some percentage breakdown of the data. The divisions into deciles (every 10 percent), quintiles (every 20 percent), and quartiles (every 25 percent) have been used quite frequently in the mapping of geographic phenomena.

The graph of a cumulative percentage distribution often is referred to as an *ogive*. Alexandersson (1956) plotted a number of these graphs for selected industries and used them as the basis for a classification of cities in the United States. The level of employment corresponding to the 5th percentile point on each graph was regarded as the level of "nonbasic" employment in the industry for each city, and deviations from this minimum level or "k" value afforded the basis for the classification. Some of the distribution curves from this study are reproduced in Figure 2.3. As might be anticipated, Alexandersson's study has been criticized for its reliance upon the somewhat arbitrary choice of the 5th percentile points as the all-important "k" values.

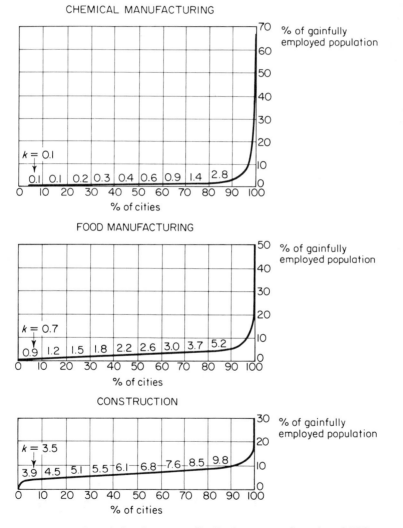

Figure 2.3. Cumulative frequency distribution curves for selected U.S. urban employment categories. *Source:* Alexandersson (1956).

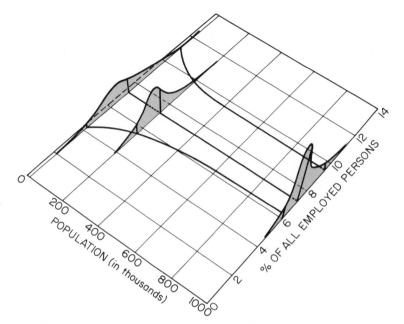

Figure 2.4. Distribution of percentages of all employed persons engaged in the construction industry in the South and West, by city size. *Note:* The lowest of the three lines in the horizontal plane represents the 5th percentiles of the conditional distributions, the middle line shows the median, and the upper line indicates the 90th percentiles. *Source:* Morrissett (1958).

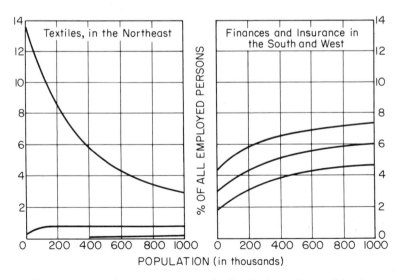

Figure 2.5. Atypical employment-ratio distributions. *Source:* Morrissett (1958).

Morrissett (1958) has extended Alexandersson's analysis to consider the effects of city size and regional location on the form of the employment distributions and the magnitude of the "k" values. One of the "employment-ratio" distributions which Morrissett derived is shown in Figure 2.4. On this graph the cumulative distribution curves for the construction industry in the South and West have been drawn for the different population-size levels. The 5th, 50th, and 90th percentile points have been plotted on the abscissa for each size level and then these points for the different sizes have been connected by curves. Two examples of "very atypical" distributions (Figure 2.5) can be contrasted with the more "typical" ones for construction.

Morrissett averaged the percentile values for some 36 industries in the different regions and plotted these mean values against population size. The examination of these composite relationships suggested to Morrissett the following generalizations:

(i)   "Small cities are much less diversified than large cities, and much more specialized."

(ii) "Cities of the South and West are more diversified, and less specialized than cities of the Northeast."

(iii) "As population increases, diversification increases and specialization decreases." Note that it is only with regard to variation in population size among the cities that the notion of population increase is defined here. No temporal process is considered.

(iv) "The proportion of persons employed in most industries is higher in large cities than in small."

(v)   "Differences in economic structures are greater between small and medium-size cities than they are between medium-size and large cities."

As in any analysis of this type, the question about the processes at work that generate these relationships remains unanswered.

Up to this point, we have been discussing the representation of a single variable X, the values of which could be plotted as points along a straight line and also grouped into a set of discrete class intervals. Another important type of data analyzed in geographic research is associated with a regional or subregional division. Now, it is of interest to identify not only each observation X_i but also the region in which the observation is located. Thus, X_{ij} is the value of X for the ith observation in the jth region. The situation is very similar to the frequency distribution for grouped data, except that a second subscript is now employed. Also, instead of referring to a frequency (f_j) for each class, we now use n_j for the number of observations in the jth region. To obtain the sum of all the X values over a set of k regions, a double summation is employed. First, the values for each region are summed:

$$Y_j = \sum_{i=1}^{n_j} X_{ij},$$

and then these regional subtotals are added together:

$$\sum_{j=1}^{k} Y_j = \sum_{j=1}^{k} \sum_{i=1}^{n_j} X_{ij}.$$

This computation is illustrated for Table 2.6. These data are the net relative changes in employment over the decade 1950–60 for the counties in Ohio. They are computed on the basis of national growth rates, county industrial mixes, and industry growth rates.

$$\sum_{j=1}^{k} Y_j = \sum_{j=1}^{k} \sum_{i=1}^{n_j} X_{ij}$$
$$= (-21.10) + (-17.80) + (-16.60) + (36.60) + (-7.60) + (-2.40)$$
$$= -28.90.$$

2.3 GEOGRAPHIC MATRICES

The data used by geographers relate to numerous phemomena measured for different areas and locations on the earth. As such, these data cannot be represented completely by way of a frequency distribution which takes account of variation in only one feature. Instead, it is useful in some problems to present the data in a two-way classification, with the locations or places on one axis and the phenomena or variables on the other.

Berry (1964) has discussed a generalization of this scheme. As a format for much of geographic research, he has suggested a "geographic matrix" in which characteristics would account for the rows, places for the columns, and each matrix cell would contain a "geographic fact." Any row of this matrix is a spatial distribution which can be mapped, and any column is "a locational inventory." The dynamics of locational arrangements and place characteristics can be represented by a number of these matrices for a series of "cross sections or slices taken through time." Haggett and Chorley (1967) develop further on this notion.

The exact specification of the rows and columns in the geographic matrix vary according to the research problem. In studies of trade and transportation, for example, the columns could be pairs (dyads) of countries or places; the rows, the various commodities; and the cells, the levels of total commodity trade between the two countries. Alternatively, a "transactions" form of the matrix would have the same places as both the rows and columns. The rows then would represent a set of origins, the columns a set of destina-

TABLE 2.6. TOTAL NET RELATIVE EMPLOYMENT CHANGES 1950–60, FOR COUNTIES IN STATE ECONOMIC SUBREGIONS OF OHIO* (DATA IN THOUSANDS)

Northeast		East-central		Central Allegheny plateau		West-central		Tri-state subregion			Flatlands
19.8	−0.5	2.3	−3.0	0.3	−1.2	30.6	0.0	1.8	−0.3	−1.6	10.2
8.5	−0.5	−0.4	−5.6	−0.2	−1.5	13.5	−0.4	1.3	−0.5	−1.9	−0.7
5.2	−0.8	−0.4		−0.6	−2.1	8.7	−0.4	0.7	−0.5	−2.1	−1.3
4.8	−3.5	−0.7		−0.7	−5.6	6.4	−0.6	0.7	−0.6		−1.8
4.7	−5.7	−1.5		−0.8		0.9	−1.0	0.4	−0.6		−1.8
2.2	−5.9	−1.5		−0.9		0.9	−2.7	−0.1	−0.6		−7.0
2.2	−5.9	−2.1		−1.0		0.7	−5.8	−0.1	−0.9		
1.2	−12.3	−2.3		−1.1		0.4	−6.8	−0.1	−1.0		
−0.4	−34.2	−2.6		−1.2		0.2	−8.0	−0.3	−1.3		

	Northeast	East-central	Central Allegheny plateau	West-central	Tri-state subregion	Flatlands
n_j	18	11	13	18	21	6
$\sum_{i=1}^{n_j} X_i$	−21.10	−17.80	−16.60	36.60	−7.60	−2.40
$\bar{X}_j$	−1.17	−1.62	−1.28	2.03	−0.35	−0.40

$$M = -0.33$$

*One subregion containing only one county is omitted.

Source: Growth Patterns in Employment by County, 1940–50 and 1950–60, Vol. 3, "Great Lakes." Washington, D.C.: U.S. Dept. of Commerce.

tions, and the cells the levels of commodity trade or migration movement between pairs of origins and destinations.

The data presented in a geographic matrix might be measured on any of the scales discussed earlier. The connectivity matrix is a form of geographic matrix with nominal measurements included. Ordinal, ratio, and interval measurements also can appear, either exclusively or in combination.

It is inevitable, given the complexity of most spatial patterns and forms, that matrix representations of geographic data will be used more frequently in geographic research in the future. As noted earlier, the matrix is a special type of number for which there is a whole body of appropriate mathematical procedures and theory. Increasingly, geographers are making use of these mathematics.

2.4 CENTRAL TENDENCY AND VARIANCE

In order to specify the overall characteristics of any frequency distribution, it is usual to consider two main features of the distribution, its *central tendency* and *variance*. A number of measures exist for each of these features and most of these have been used quite frequently in geographic research. We review here only the more conventional applications; the discussion of central tendency and variance in two-dimensional patterns is left for a later chapter.

With regard to central tendency, the crudest measure is the *mode*, which is defined simply as the most frequently occurring value. In the case of grouped data, however, it may be possible to identify only a modal class, and interpolation is needed to obtain a single value for the mode. Furthermore, frequency distributions often have more than one mode, the bimodal case being quite common.

A better measure of central tendency is the *median*, which is the middle value in a ranking of the complete distribution of values. An interesting application of this statistic is suggested in the following discussion by Alonso (1964, pp. 79–80) of a very simple location problem. Consider

> The location of a firm which, let us say, makes and delivers bakery products. Neither the cost of making these products nor the volume of business will vary with the location of the firm. The only variable in this case is the delivery costs. The customers, A, B, ... G, each take one delivery a day, and are distributed along a road as shown in Figure 2.6. The bakery sends out a boy who can carry only one customer's order at a time, so that he has to make one trip per customer. Where then to locate the bakery to minimize the boy's trips? The almost automatic answer would be the "average," center of gravity, or mean location. This is easily found by summing the distances from either end and dividing by the number of customers. In this

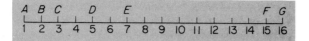

Figure 2.6. Distribution of the bakery's customers. *Source:* Alonso (1964).

case, summing from A, it would be $0 + 1 + 2 + 4 + 6 + 14 + 15 = 42$; dividing by the number of customers or trips (7), the mean is 6 blocks to the right of A, at the same location as E. But this is the wrong solution. Examine Table 2.7. The total distance is less if the bakery locates at D than if the bakery locates at E. If we had gone about the problem in a systematic fashion, we should have asked: which location minimizes the sum of the distances from the bakery to its customers? This can readily be solved by elementary calculus. In fact, however, we might have recognized that the point on a distribution along a line at which the total distance to all other points is minimized is the median (that is to say, the point at which there are as many points to one side as to the other). The median in this case is D.

TABLE 2.7.   TOTAL TRIPS ACCORDING TO LOCATION OF THE
BAKERY AT E OR D IN FIGURE 2.6

Customer	Distance from location at E	Distance from location at D
A	6	4
B	5	3
C	4	2
D	2	0
E	0	2
F	8	10
G	9	11
Total distance traveled	34	32

Although the location problem posed in this example is simple, it does illustrate that for any variable X the sum of the absolute deviations of the X_i values from some point A is minimized when A is the median. That is to say, the expression

$$\sum_{i=1}^{N} |X_i - A|$$

is minimized when A is the median value for the set of X_i's.

Again, for grouped data, interpolation is required to obtain the median value. Similarly, with ungrouped data, when the total number of observations is even, say 10, it may be desirable to interpolate a value between the two middle values, that is, between the fifth and sixth, as the median. In any cumulative percentage distribution the median corresponds to the 50th percentile point.

In the sense that the median is influenced only by the position of values in a ranking and not directly by the magnitude of these values, it is an appropriate measure of central tendency in skewed distributions. In this respect, it contrasts with the other more powerful and more commonly used measure of central tendency, the *mean*, which is computed from the actual sum of the values.

For ungrouped data, the mean ($\bar{X}$) is given by the expression

$$\bar{X} = \left(\sum_{i=1}^{N} X_i \right) \Big/ N;$$

whereas for grouped data, it is given by

$$\bar{X} = \left(\sum_{j=1}^{k} f_j \hat{X}_j \right) \Big/ \sum_{j=1}^{k} f_j,$$
$$= \left(\sum_{j=1}^{k} f_j \hat{X}_j \right) \Big/ N,$$

where k is the number of classes or groups, f_j is the associated class frequency, and $\hat{X}_j$ is the class midpoint value.

For the data given in Tables 2.4 and 2.5 the mean values are as follows.

(i)   Ungrouped:
$$\bar{X} = 3860/157 = 24.59.$$

(ii) Grouped:
$$\bar{X} = 3996.5/157 = 25.46.$$

The mean is a relatively poor measure of central tendency for skewed distributions, reflecting as it does the weighting of the extreme values in the "tail" of the distribution. Unfortunately, this fact is not always kept in mind, and as we shall note shortly, the use of the mean as the "norm" from which to measure deviations has often produced disappointing results in such problems as city classification.

It was mentioned that the mean is a powerful measure of central tendency. It is in the sense that it features prominently in many of the theorems of mathematical statistics and has a formal role in probability theory. Later, we shall refer to the mean as the "expectation" of a probability distribution. For the time being, however, we shall concern ourselves with some other features of the *mean*.

The mean, in a sense, is the center of gravity for the set of values in a frequency distribution. Along with some other measures of a frequency distribution, it often is referred to as a *moment* of the distribution, the term having been borrowed from mechanics. In statistical analysis, considerable use is made of the moments about zero, including the mean, and also of the moments about the mean itself. We shall discuss shortly the *variance* as one of the important moments about the mean. At this point we note that any moment (M_p) about the mean is given by the expression

$$M_p = \left[\sum_{i=1}^{N} (X_i - \bar{X})^p \right] \Big/ n,$$

where n, as we shall see, represents the appropriate degrees of freedom.

With respect to the data given in Table 2.4, it is possible to verify the following statements involving the mean:

(i) $\left[\sum_{i=1}^{N} (X_i + k) \right] \Big/ N = \bar{X} + k,$ where k is any constant.

(ii) $\left(\sum_{i=1}^{N} kX_i \right) \Big/ N = k\bar{X},$ where k is any constant.

(iii) $\sum_{i=1}^{N} (X_i - \bar{X}) = 0;$ that is, the first moment about the mean is zero.

(iv) $\sum_{i=1}^{N} (X_i - A)^2$ is minimized when A is the mean $\bar{X}$.

(v) $\sum_{i=1}^{N} (X_i - \bar{X})^2 = \sum_{i=1}^{N} X_i^2 - \left(\sum_{i=1}^{N} X_i \right)^2 \Big/ N.$

The first three identities are proved easily, demanding only some simple manipulations with the summation operator. The fourth statement is important, for it enables a minimum expression to be defined for the variation within a set of values. The proof of the theorem involves the calculus and is not presented here. The fifth statement provides a convenient computational formula for the second moment about the mean, which we shall take up shortly.

There are many important variations on the concept of a mean. In some situations, for example, it may be desirable to give greater weight to some values than to others, and a *weighted mean* may be computed as

$$\left(\sum_{i=1}^{N} w_i X_i \right) \Big/ \sum_{i=1}^{N} w_i,$$

where w_i is the weight assigned to the ith observation. Or in averaging ratios or rates of change, it is usually preferable to calculate the *geometric mean*, which is defined as

$$\left(\sum_{i=1}^{N} \log X_i \right) \Big/ N.$$

There are other special types of averages besides these two, but this topic is not pursued here.

A measure of central tendency tells us something about a frequency distribution but we usually are interested also in knowing how the values are distributed around this central one. This variation can be measured in a number of ways.

The simplest measure of variation is the *range* between the largest and smallest values. This is a fairly crude measure and ignores too much of the available information to be of very great interest. Similarly, the *semi-interquartile range* (*Q*), defined as $(Q_3 - Q_1)/2$, where Q_3 and Q_1 are the 75th and 25th percentile points, respectively, provides only limited information concerning the variability of a set of data. The basic problem with both of the preceding measures is that they ignore most of the values and focus attention only upon two specific values.

By contrast, the *mean deviation* (*MD*) considers all the values in a distribution and is an average of the absolute deviations of these values from some point, usually the median or mean. When the median is used, then

$$MD = \left(\sum_{i=1}^{N} |X_i - Mdn|\right)\bigg/ N,$$

and we have noted already that the sum in the numerator of this equation is the minimum of the expression

$$\sum_{i=1}^{N} |X_i - A|,$$

where A is any point in the distribution. Unfortunately, the absolute values in this measure of variation do not allow for easy mathematical manipulation. Therefore, greater reliance has been placed in statistical analysis upon the *mean square*, which in its general form is given as

$$\left[\sum_{i=1}^{N} (X_i - A)^2\right]\bigg/ N,$$

where A is again any point in the distribution. It is usual for A to be taken as the mean ($\bar{X}$), and in this case the mean square is simply the second moment about the mean or, as it is commonly referred to, the *variance*. In writing an expression for the variance, we shall make one small change in the formula given above. The denominator will be $(N - 1)$ and not N. Hence, the variance (s^2) is

$$s^2 = \left[\sum_{i=1}^{N} (X_i - \bar{X})^2\right]\bigg/ (N - 1).$$

The modification deals with quite a sophisticated concept in mathematical statistics, that of the "degrees of freedom." We already know that

$$\sum_{i=1}^{N} (X_i - \bar{X}) = 0,$$

so that in computing the numerator of the variance once we have determined $(N - 1)$ of the deviations $(X_i - \bar{X})$, the Nth one is automatically fixed since the sum of these first-order differences must be zero. We are, of course, free to compute first whichever $(N-1)$ values we choose and for this reason there are said to be $(N-1)$ degrees of freedom in this situation. Therefore, the denominator of the expression for the variance is usually given as $(N-1)$, and we shall follow this practice from now on.

The concept of variance is fundamental to statistical analysis. Unless it is otherwise qualified in this book, the term "variation" always will be defined in terms of variation from the mean, and the variance will be assumed to be the measure of this variation.

The square root of the variance is the *standard deviation* (*s*), and this statistic has been used frequently in geographic research. One such application is reviewed briefly. In his classification of 897 urban concentrations in the United States, Nelson (1955) computed means and standard deviations for nine major employment categories. The urban centers were then classified according to whether their employment levels in these categories were within one, two, three, or more standard deviations *above* the mean levels. The frequency distributions for two of the employment categories are shown in Figure 2.7. Unfortunately, these diagrams emphasize some of the weaknesses in the approach which Nelson adopted. Consider the expression for the standard deviation

$$s = \left\{ \left[\sum_{i=1}^{N} (X_i - \bar{X})^2 \right] \Big/ (N-1) \right\}^{1/2}$$

and the fact that it involves a sum of squared values in the numerator. This statistic is unaffected by the direction of deviations from the mean, but it is affected by the magnitude of the deviations. Therefore, in the case of skewed distributions the standard deviation will be weighted considerably by the extreme values. This suggests that the standard deviation is a more appropriate measure for symmetric distributions, and in situations such as in Figure 2.7 its use gives rise to difficulties. Nelson's distributions are not symmetric and they are truncated at zero. Even if Nelson had chosen to consider deviations below the mean, which he did not, it is obvious that his class limits would have involved negative percentages of employment, which are meaningless.

The expression for the variance often is given as

$$s^2 = \left(\sum_{i=1}^{N} x_i^2 \right) \Big/ (N-1),$$

where it is understood that

$$\sum_{i=1}^{N} x_i^2 = \sum_{i=1}^{N} (X_i - \bar{X})^2 = \sum_{i=1}^{N} X_i^2 - \left[\left(\sum_{i=1}^{N} X_i \right)^2 \Big/ N \right].$$

The representation of data in *deviation form* is quite common in statistical analysis, and we shall have occasion later to make frequent use of this representation.

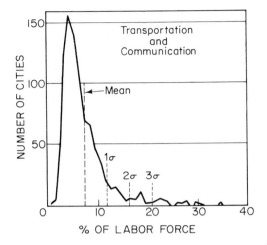

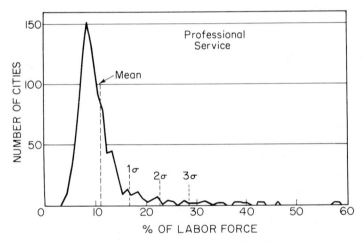

Figure 2.7. Cumulative frequency curves for selected employment categories in U.S. cities. *Source:* Nelson (1955).

For the variance we note that in the cases where k is any constant,

$$s^2_{(X+k)} = s^2_X,$$

and

$$s^2_{kX} = k^2 s^2_X.$$

For grouped data the variance is given by the expression

$$s^2 = \left[\sum_{j=1}^{k} f_j(\hat{X}_j - \bar{X})^2\right]\Big/(N - 1).$$

The computations for the data in Tables 2.4 and 2.5 are given below.

(i)   Ungrouped:

$$s^2 = \left[\sum_{i=1}^{N} X_i^2 - \left(\sum_{i=1}^{N} X_i\right)^2 \bigg/ N\right]\bigg/(N-1)$$
$$= \left[139962.00 - \frac{(3860)^2}{157}\right]\bigg/(157-1)$$
$$= (139962.00 - 94901.91)/156$$
$$= 45060.09/156$$
$$= 288.85$$
$$s = \sqrt{288.85}$$
$$= 16.99.$$

(ii)   Grouped:

$$s^2 = \left[\sum_{j=1}^{k} f_j(\hat{X}_j - \bar{X})^2\right]\bigg/(N-1)$$
$$= 42756.44/156$$
$$= 274.08$$
$$s = 16.55.$$

All the measures of variation discussed so far are expressed in the same metric units as the original data. This precludes meaningful comparison of the variability of several frequency distributions if different scales of measurement are involved. To overcome this difficulty, the *coefficient of variation*, defined as the ratio of the standard deviation to the mean, is often employed. Fuchs (1960), in his study of the variability in U.S. urban residential standards, made use of this coefficient. In this study, emphasis was placed on the variation in housing costs among 209 cities of over 50,000 population in 1950. The cities were divided into two groups: central cities of metropolitan areas and noncentral cities. Then for the cities in each group the coefficients of variation (c.v.) in housing costs were computed. Some of the results are summarized in Table 2.8. On the basis of these results,

TABLE 2.8.   COEFFICIENTS OF VARIATION IN HOUSING COSTS

Central cities		Noncentral cities
163	number of cities	46
54.0–117.5	range of c.v.	43.6–89.8
75.4	mean c.v.	64.9
71.1	median c.v.	62.1

Source: Fuchs (1960, p. 317).

Fuchs concluded that "the noncentral cities are generally less varied and more uniform in residential quality than central cities." An attempt then was made to explain the differences among the cities in the coefficients of variation in terms of selected income and housing characteristics.

Before leaving the topic of variance, it will be useful for later purposes to consider briefly the question of measuring variation for the regional data set introduced earlier.

Recall the situation in which there are N observations located in k different regions. How is total variation to be defined now? In this case, it can be shown that the appropriate sum of squares is given as

$$\sum_{j=1}^{k} \sum_{i=1}^{n_j} (X_{ij} - M)^2 = \sum_{j=1}^{k} \sum_{i=1}^{n_j} (X_{ij} - \bar{X}_j)^2 + \sum_{j=1}^{k} n_j (\bar{X}_j - M)^2,$$

where

X_{ij} is the ith observation in the jth region,
$\bar{X}_j$ is the mean value of X for the jth region,
n_j is the number of observations in the jth region,
M is the grand mean of all the X values.

The expression above is an important one, for as we shall see, it is basic to one of the more powerful tools of statistical analysis, the analysis of variance. The expression states that the sum of the squared deviations of the observation values from the grand mean, the sum of squares for total, can be partitioned into two parts, the sum of the squared deviations of the values from their respective regional means and the sum of the weighted squared deviations of the regional means from the grand mean. The corresponding expressions for the data given earlier in Table 2.6 are shown below.

$$\sum_{j=1}^{k} \sum_{i=1}^{n_j} (X_{ij} - M)^2 = \sum_{j=1}^{6} \sum_{i=1}^{n_j} (X_{ij} - \bar{X}_j)^2 + \sum_{j=1}^{6} n_j (\bar{X}_j - M)^2$$
$$= (1959.44 + 38.62 + 24.42 + 1316.00 + 19.19 + 160.74)$$
$$+ (12.70 + 18.31 + 11.73 + 100.25 + 0.01 + 0.03)$$
$$= 3518.41 + 143.03$$
$$= 3661.44.$$

The concepts of central tendency and variance introduced in this chapter are basic for much of the discussion in the following chapters. So far, however, we have developed the concepts only with respect to one variable measured on a single scale. We shall have need later to generalize them for handling two other situations: first, the case of a variable measured in two dimensions, that is, a geographic location pattern; and second, the analysis of the joint variation among two or more variables. But before these situations are considered, we have to review some other important topics concerning the univariate case. Foremost among these is the theory of probability developed with respect to a single random variable. This topic is taken up in the next chapter.

SUGGESTED READINGS

Berry, B. J. L. (1964), "Approaches to Regional Analysis: A Synthesis." *Annals*, Association of American Geographers, Vol. 54, pp. 2–11.

Campbell, N. (1952), *What is Science?* New York: Dover Publications, Inc., Chapters 6–8.

Duncan, O. D., et al. (1961), *Statistical Geography: Problems in Analyzing Areal Data*. New York: The Free Press, Chapters 1–2.

Fuchs, R. J. (1960), "Intraurban Variation of Residential Quality." *Economic Geography*, Vol. 36, pp. 313–325.

Garrison, W. L. (1956a), "Some Confusing Aspects of Common Measurements." *The Professional Geographer*, Vol. 8, pp. 4–5.

Hodge, G. (1963), "Use and Mis-use of Measurement Scales in City Planning." *Journal of American Institute of Planners*, Vol. 29, pp. 112–121.

Isard, W. (1960), *Methods of Regional Analysis: An Introduction to Regional Science*. New York: John Wiley & Sons, Inc., Chapter 7.

Morrissett, I. (1958), "The Economic Structure of American Cities." *Papers*, The Regional Science Association, Vol. 4, pp. 239–258.

Wright, J. K. (1937), "Some Measures of Distributions." *Annals*, Association of American Geographers, Vol. 27, pp. 177–211.

PROBABILITY FUNCTIONS
IN GEOGRAPHIC RESEARCH

Probability theory provides the mathematical foundations upon which statistical analysis is built. Introductory statements of the theory are available in several excellent books such as Goldberg (1960); Kemeny, et al. (1966); Parzen (1960); and Hodges and Lehmann (1965). No attempt is made here to outline this theory, but as a background for the review of the applications of probability models in geographic research which is the subject of this chapter particularly, it is important to note some of the more important concepts and definitions.

3.1 SOME FUNDAMENTAL CONCEPTS

3

Probability theory has developed from formal consideration of random phenomena, that is to say, from the study of phenomena which on any single observation may result in any one of a number of possible outcomes. The particular outcome on a single observation is dependent on chance and cannot be predicted precisely ahead of time. It is understood, however, that over repeated observations the outcomes will show some statistical regularity in the sense that numbers between 0 and 1 can be used to represent the relative frequencies with which the different possible outcomes occur. The total collection of all possible outcomes constitutes the *sample space S*, whereas any one of the outcomes is described as an *event E*. Given a very large number of observations, the relative frequency of a random event will approach a stable or fixed value, which is called the *probability* of that event.

Introductory discussions of sample spaces and events usually are couched in the language of the algebra of sets. For illustrations of some basic set theory notions, consider the diagram Figure 3.1. Here is a mapping of the sample space for the random phenomenon represented by the rolling of two unbiased dice. Each point in the diagram represents an outcome or event. We can define events which may be represented by several of the points in the sample space. For example, consider the event that "the sum of the two numbers showing after the dice have been thrown is 7." Then this event, call it E_1, involves the six sample points (6, 1), (5, 2), (4, 3), (3, 4), (2, 5), and (1, 6). Similarly, the event E_2 that "the number showing on die B is a 3, 4, or 5," involves the 18 points shown on the diagram. Now the *intersection* of the two events E_1 and E_2 is defined as the collection of sample points which belong to both subsets, in this case (2, 5), (4, 3), and (3, 4). The intersection of two events is written as $E_1 \cap E_2$, and this concept can be extended to the case of more than two events. The *union* of the two events involves

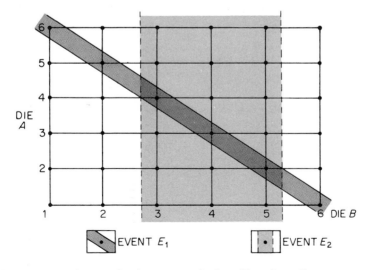

Figure 3.1. Sample space for the outcomes in the rolling of two dice.

all the sample points, in this case 21 of them, which belong to at least one of the two events. In the language of sets, the union is written as $E_1 \cup E_2$, and again the concept can be generalized for more than two events. Two events are said to be *mutually exclusive* if their intersection is impossible or does not take place. This would be the case, for example, with the event E_2 as defined above and another event E_3, defined as "the sum of the two dice faces is 3." In more formal discussions of set theory these concepts often are developed with reference to diagrams known as *Venn* diagrams.

Some geographers, notably Minnick (1964), Cole (1966), Golledge and Amadeo (1966), and Rodoman (1967) have used set theory in their discussions of regions, subregions, and spatial association. Minnick, for example, suggests a "coefficient of areal correspondence" defined as the ratio $(A \cap B)/(A \cup B)$, which is interpreted as "where either region A or B, there region $A \cap B$." Shear (1966) has attempted to formulate the Koppen classification of dry climates in set theoretic terms.

To the extent that this exploratory work in set theory is suggestive of a new conceptual framework for geographic research, then it is worthy of note. At this point, however, it appears somewhat premature and evades the main issue, that of the need for the geographer to identify clearly the random phenomena in which he is interested, the nature of the sample spaces, and the probabilities associated with the different outcomes. It is this last point, stating the probabilities of events, that we now discuss.

Consider again Figure 3.1. For this sample space S, we seek a rule which will assign to every event E in the sample space a probability P such that at least three conditions are satisfied.

(i) $P(E) \geqslant 0$ for every event E. In other words, an event cannot have a negative probability.

(ii) $P(S) = 1$ for the certain event S, that is, the whole sample space. An impossible event $\emptyset$, for example, that the sum of the two numbers showing on the dice is 13, has a probability of zero.

(iii)   Finally, if E_2 and E_3 are mutually exclusive, then $P(E_2 \cup E_3) = P(E_2) + P(E_3)$. If two events are not mutually exclusive, then $P(E_1 \cup E_2) = P(E_1) + P(E_2) - P(E_1 \cap E_2)$.

In the example of Figure 3.1 these requirements are met if the probability of an event is given as *the relative frequency of the sample points involved in the event to the total number of sample points.* Hence, the probability of the event E_1, that the sum is 7, is

$$P(7) = \tfrac{6}{36} = \tfrac{1}{6}.$$

Alternatively, if each sample point is considered as an event with a probability of $\tfrac{1}{36}$, then the probability of the compound event E_1 is

$$P(7) = P(6, 1) + P(5, 2) + P(4, 3) + P(3, 4) + P(2, 5) + P(1, 6)$$
$$= \tfrac{1}{36} + \tfrac{1}{36} + \tfrac{1}{36} + \tfrac{1}{36} + \tfrac{1}{36} + \tfrac{1}{36}$$
$$= \tfrac{1}{6},$$

since these individual events are mutually exclusive.

In geography, very little is known about the probability rules which are appropriate in dealing with spatial events. It is only in physical geography, particularly climatology and more recently in the analysis of spatial point patterns, that much attention has been given to this question.

One of the problems in dealing with spatial events is that they are seldom independent of one another. That is to say, the probability of one event often is conditional upon the occurrence of some other event. The conditional probability of an event E_1, given that another event E_2 has already occurred, is written as $P(E_1 | E_2)$. In our earlier example, the probability of a 7 total, given that a 3, 4, or 5 occurs on die B, is the sum of the probabilities of the three events (2, 5), (4, 3) and (3, 4) in the subspace associated with E_2. Hence

$$P(E_1 | E_2) = P(7 | 3, 4, \text{ or } 5) = P[(2, 5) | E_2] + P[(4, 3) | E_2] + P[(3, 4) | E_2]$$
$$= \tfrac{1}{18} + \tfrac{1}{18} + \tfrac{1}{18}$$
$$= \tfrac{1}{6}.$$

This illustrates the point that

$$P(E_1 | E_2) = \frac{P(E_1 \cap E_2)}{P(E_2)}, \qquad \text{for } P(E_2) > 0.$$

From this, definitions of *independent* and *dependent* events follow. Two events defined in the same sample space S are said to be independent if the following relation holds:

$$P(E_1 | E_2) = P(E_1),$$

which implies that

$$P(E_1 \cap E_2) = P(E_1) \cdot P(E_2).$$

This rule for the multiplication of the probabilities of *independent* events to obtain the probability of their intersection is important. If the relations above do not hold, then the events are said to be *dependent*. Again, these definitions can be extended to the case of more than two events.

The discussion up to this point has been in terms of any random phenomenon for which the sample space of outcomes is finite and completely specified. Probability theory, however, is developed in far more general terms. Consider a *random variable* X, which takes on numerical values $(x_1, x_2, x_3, \ldots, x_n)$ from among the set of all real numbers $-\infty$ to $+\infty$. What forms of probability rules can be used in assigning probabilities to these outcomes?

There are two different types of mathematical functions which serve to specify these rules. The first we shall refer to as *probability mass* or *density* functions; the second as *distribution* functions. What is the distinction?

(i) For many probability rules there exists a mathematical function $f(x)$, defined for real numbers x, which gives the probability for any particular outcome. In many cases, $f(x)$ may be zero except for a limited set of x values at which the function is positive. Consider, for example, the probabilities associated with the different outcomes relating to the number of heads obtained when an unbiased coin is tossed 10 times. The possible outcomes are

$$x = 0, 1, 2, 3, 4, 5, 6\ 7, 8, 9, 10$$

and the probabilities associated with these are given by

$$p(x) = \frac{10!}{x!(10 - x)!}(0.5)^x(0.5)^{10-x}.$$

We shall refer to this function later as the *binomial function*, and at this time we note that at all points other than $(x = 0, 1, 2, 3, 4, 5, 6, 7, 8, 9, 10)$ the probability is zero (Figure 3.2). Any such function which is positive only for a finite or at least countably infinite set of values is called a *probability mass function*.

In contrast to the situation just outlined, $f(x)$ often will be defined for all real numbers, as shown in Figure 3.3. The probability of an event now is given as an area under a curve, and the integral calculus must be used to find this value. That is to say,

$$P(E) = \int_E f(x)\, dx.$$

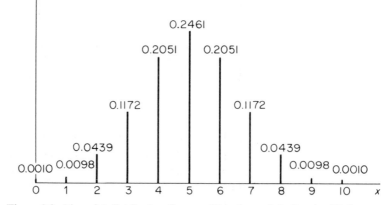

Figure 3.2. Binomial distribution for $n = 10$ and $p = 0.5$. *Source:* Mode (1966).

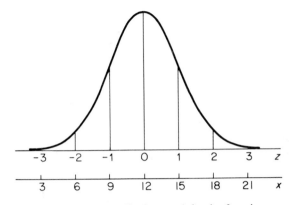

Figure 3.3. Standardized normal density function.

Thus in Figure 3.3 the probability of the event that x is between 10 and 20 would be given as

$$\int_{10}^{20} f(x)\, dx.$$

In this case, $f(x)$ is a *probability density function*, and it follows that

$$\int_{-\infty}^{+\infty} f(x)\, dx = 1.$$

(ii) Now, it is possible to define another function of x which alone will yield all the required information about the probability rule. This new function is the *distribution function* $F(x)$, and for any real number x it gives the probability that an observed real value x' of the random variable is less than or equal to x. Thus for any real number x, we can write this as

$$F(x) = P[(\text{for a real number } x': \quad x' \leqslant x)].$$

The connection with the cumulative percentage distribution which was discussed earlier should be clear.

In the case of a probability mass function,

$$F(x) = \sum_{k} f(x'), \qquad (3.1)$$

where k represents the set of points $x' \leqslant x$, which satisfies the relation $p(x') > 0$. If on the other hand a probability density function is involved, then

$$F(x) = \int_{-\infty}^{x} f(x')\, dx'. \qquad (3.2)$$

A random variable which has a distribution function of the form given in equation (3.1) is called a *discrete* variable, whereas a distribution function of the form (3.2) is called *continuous*. The graphs of examples of these types of distributions are shown in Figure 3.4. It is, of course, possible to define

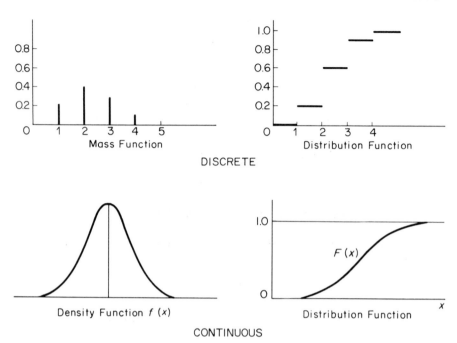

Figure 3.4. Continuous and discrete probability distributions.

a mixed distribution function which combines both types of distributions. In many research problems, phenomena which logically appear discrete, for example, the number of employees, factories, or urban centers in an area, often are treated as continuous functions in order to facilitate the analysis.

There are some other important concepts to be introduced concerning a random variable. That of the *moments* of a variable, which was introduced in the preceding chapter, is defined now in terms of mathematical *expectation*. For a random variable X, the mean is denoted by $E[X]$, the variance by $E[(X - E[X])^2]$, and the nth moment about the mean by $E[(X - E[X])^n]$. In the case of a discrete distribution,

$$E[X] = \sum_{k=1}^{n} x_k f(x_k)$$

and

$$E[(X - E[X])^2] = E[X^2] - (E[X])^2$$

$$= \sum_{k=1}^{n} x_k^2 f(x_k) - \left[\sum_{k=1}^{n} x_k f(x_k)\right]^2.$$

For a continuous distribution,

$$E[X] = \int_{-\infty}^{+\infty} x f(x)\, dx$$

and

$$E[(X - E[X])^2] = \int_{-\infty}^{+\infty} x^2 f(x)\, dx - \left[\int_{-\infty}^{+\infty} x f(x)\, dx \right]^2.$$

The reader is referred to Goldberg (1960, pp. 172–181) and Parzen (1960, pp. 342–390) for lengthier and more formal discussions of these concepts.

In many cases the computation of the various moments for a probability distribution is facilitated by use of a special mathematical expectation, the *moment-generating function*. Not all probability distributions have such functions, and in some situations it is necessary to rely upon another function, *the characteristic function*, which always exists, in order to determine the probability distribution. Again, the derivation of these functions involves mathematics of a higher order than is attempted in this book, and the reader should pursue these topics in more formal texts, for example, Parzen (1960, pp. 391–413). They are mentioned here because in one or two instances in the geographic literature the functions have been used in specifying the characteristics of probability distributions applicable to the locational patterns of geographic phenomena.

3.2 APPLICATIONS OF PROBABILITY MODELS IN GEOGRAPHIC PROBLEMS: ONE-DIMENSIONAL SITUATIONS

There have been only a few published studies in geography which have attempted to make use of some of the basic probability distributions. The trend in the quantitative research, however, is toward an increasing emphasis on probability models. This section reviews applications which relate to one-dimensional problems. In this context, a probability distribution is derived for a variable which is measured along a single linear scale. The discussion of probability distributions for two-dimensional situations is reserved for the next section.

It is appropriate at the outset to enter a general qualification about many of the applications cited in this section. In the introductory review of basic probability concepts it was noted that the probabilities for the total set of outcomes in any sample space must sum to *one*. In the case of continuous variables, this same requirement is expressed in terms of the total area under the curve being equal to *one*. In many of the studies mentioned below, however, this requirement seemingly is not satisfied. This stems from the fact that for illustration purposes we shall view studies in a somewhat more formal way than the authors perhaps intended. We sometimes shall consider as probability laws, functions which the authors used merely as convenient descriptions of relationships.

The applications of some discrete distributions are discussed first.

The Binomial Distribution

Consider the map of a fairly large region divided into quadrats which are very small in comparison to the total area. The occurrence of forest throughout the region is assumed to be a random phenomenon, that is to say, every quadrat has an equal probability of being forested, and the character of any quadrat is assumed to be quite independent of the character of any other. Neither of these assumptions is realistic but we shall insist upon them for the present time. It is assumed further that the whole region is known to be half-forested; in other words, one-half of the quadrats are forested. Therefore, the probability p that any quadrat is forested can be taken to be 0.5; the probability q that it is not forested is therefore $(1 - p) = 0.5$ also.

Now assume that N of the quadrats are to be selected randomly. What is the probability associated with the event that exactly r of the N quadrats chosen will be forested ones?

It can be shown that for any N objects there are

$$\frac{N!}{r!(N-r)!}$$

possible combinations of exactly r objects. This topic in counting is discussed in most introductory mathematical texts, for example, Kemeny, et al. (1966, pp. 84–108). We note that

$$\frac{N!}{r!(N-r)!}$$

is written usually as $\binom{N}{r}$. Also, it follows that $\binom{N}{r} = \binom{N}{N-r}$.

The probability associated with any one of these combinations is $p^r q^{N-r}$, so that for the total number of combinations the probability will be

$$\binom{N}{r} p^r q^{N-r} = \frac{N!}{r!(N-r)!} p^r q^{N-r} \qquad \text{for all } r = 0, 1, \ldots, N,$$
$$= 0 \qquad\qquad\qquad\qquad \text{otherwise.}$$

This probability mass function is known as the *binomial probability law*. The mean of this function is equal to Np; the variance is Npq.

The binomial law has not proved useful in geographic research. The fact that for most spatial patterns the probability associated with any locational event usually is fairly small means that another discrete distribution, the Poisson, is more appropriate. It is seldom the case with locational patterns that both p and q will be around 0.5, and the more likely possibility is that one of the probabilities will be extremely low, say 0.01. In these situations, it can be shown that the Poisson serves as a limiting form of the binomial.

The Poisson Distribution

With regard to a set of quadrats such as mentioned above, assume there are a number of points distributed throughout the region with a mean density of m points per quadrat. In Figure 3.5 this density is 1.37. Assume further that the distribution of these points is random in the sense that every quadrat has an equal chance of receiving a point, and every point in turn has an equal chance of occurring in any one quadrat. Also, the location of any point is considered to be independent of the location of any other. We now ask, What is the probability that a quadrat will contain exactly x points?

Early work by plant ecologists, notably Clark and Evans (1954) on

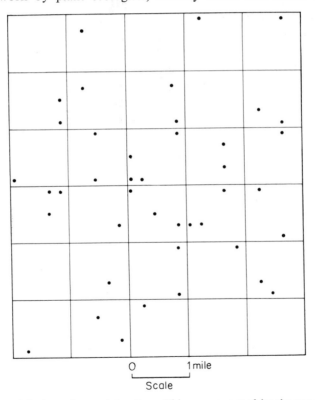

Figure 3.5. A random point pattern. This was generated by the use of a finer grid overlay and a random numbers table.

the location pattern of plants within an area, assumed that this probability is given by the Poisson function

$$p(x) = \frac{e^{-\lambda}\lambda^x}{x!} \quad \text{for } x = 0, 1, \ldots, \text{ and } \lambda > 0,$$

$$= 0 \quad \text{otherwise.}$$

The estimator of λ is the empirical density figure m. Both the mean and variance of this function are equal to λ.

Dacey (1964a) has given the formal derivation of the Poisson probability law for a two-dimensional random point process. The derivation is based on a set of four postulates which are stated as follows:

(i) *Statistical equilibrium.* Given a large uniform plane with a density of λ points per unit area and a region A selected from this plane, then the probability that a subregion of A contains exactly x points is the same regardless of the shape and location of A in the plane.

(ii) *Independence of events.* The probability that one point occurs in any subregion of A is λdA, and the probability that more than one point occurs in dA is of smaller magnitude.

(iii) *Differentiability with respect to A.* For all values of A, the probability that x points occur in a region of area A has a continuous derivative with respect to A.

(iv) *Boundary conditions.* If $P(x, A)$ is the probability that A contains exactly x points, and assuming A is not negative and $x \neq 0$, then

$$P(-x, A) = 0$$
$$P(0, 0) = 1$$
$$P(x, 0) = 0.$$

As an empirical example of the use of this probability function, consider Figure 3.5. The point pattern in this map is hypothetical, but it serves to illustrate the use of the model. The Poisson frequencies of quadrats for the point pattern shown in Figure 3.5 are given in Table 3.1. These expected frequencies agree fairly closely with the observed ones. We shall leave for the

TABLE 3.1.   DISTRIBUTION OF POINTS BY QUADRAT IN FIGURE 3.5

x	Observed frequencies	Poisson frequencies $\lambda = 1.37$
0	6	7.8
1	12	10.5
2	8	7.1
3	3	3.2
4	1	1.1

next chapter the question of deciding what is an acceptable "fit" in these circumstances.

Inasmuch as the Poisson law, when it is applied to a location pattern, implies that the process generating the pattern is a random one, then it provides a useful norm against which empirical patterns can be compared. Some examples of the use of the Poisson model in geographic research are mentioned below.

Getis (1964), in a study of the location of grocery stores in East Lansing, used the Poisson law as a theoretical model of the spatial pattern. After reviewing the historical trends affecting retail store location, Getis formulated a hypothesis as follows. "Between a period having a grouped pattern (few supermarkets) and a period having a dispersed or uniform pattern (many supermarkets), one would expect to find a random pattern, that is, there would be a pattern reflecting neither extreme population densities nor very modern densities." One of the tests of this hypothesis involved a division of the city area into quadrats and the construction of frequency distributions of the quadrats according to the number of stores they contained. These frequency distributions for 1900, 1910, and 1960 were very similar in form to the expected distribution of quadrats given by the Poisson model, and in this sense the hypothesis was substantiated. Further reference to this study will be made later.

In another approach to the problem of testing this same hypothesis, Getis examined the average distance separating stores from their nearest neighboring stores. This observed mean distance value was compared then with an expected mean distance for a random distribution of stores described by a Poisson probability model. The use of the Poisson model in this context was only an implicit one, and the same approach was adopted by King (1962a) in a study of the spacing of urban settlements in selected areas of the United States.

These two studies serve to point up some of the difficulties in the application of probability models to such research problems. Existing location theory does not lend strong support to the notion that either grocery store or city locations are random and independent processes. On the contrary, it is known that the probability of a store locating in any area is conditional upon a number of factors, not the least important of which is the relative location of other stores. Therefore, a simple model such as the Poisson law hardly is suggested by theory, and while it may serve as a convenient first approximation of the location pattern, it reduces to simplicity a situation which already has been acknowledged as a complex one. Besides, it is almost certain that other probability models could be found which would fit the observed facts equally as well, and unless there is theory to guide us in our choice, one model may appear no better than the others.

Curry (1962) used the Poisson model in describing the probability distributions for trip distances associated with different types of urban goods and services. The emphasis in this case was on the distribution of events over time intervals rather than spatial units. Some of the Poisson curves used by Curry are shown in Figure 3.6. These are for central place goods which differ according to the mean number of trips made per week to obtain them. Implicit in Curry's discussion is an interesting approach to the question of using probability models in geographic research. In his study, Curry was not concerned with any empirical testing of how well the Poisson model fitted the data. Rather, he used these functions as inputs to a more general model and proceeded on the assumption that the Poisson was an adequate description of certain aspects of reality. Viewed in this light, the criticisms noted in the preceding paragraph with regard to applications of the Pois-

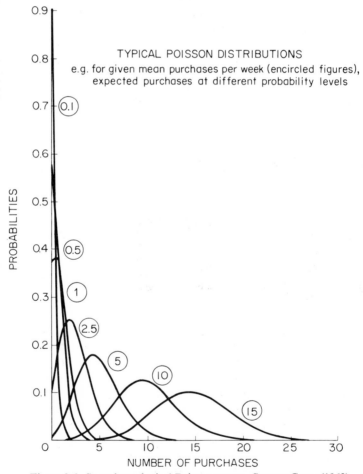

Figure 3.6. Some hypothetical Poisson curves. *Source:* Curry (1962).

son model may not appear so serious. The fact that there are no obvious correspondence rules between the generating processes for a probability model and what is known about reality may sometimes be unimportant as long as the model provides a reasonably good description of the reality and can serve as an input to a more complex study. Many studies in queuing theory and attempts to develop simulation models appear to proceed on this assumption.

Negative Binomial Law

Different authors have drawn attention to at least two serious interrelated problems associated with the use of the Poisson law in analyzing geographic patterns. The first is that the mathematical assumption concerning the independence of events is seldom appropriate for the problems in which the geographer is interested. This point has been made already with respect to the studies by Getis and King. Related to this is the second problem that there frequently is a tendency in location patterns toward either regular spacing or clustering. To handle this problem, plant ecologists and others have made use of certain "contagious" models in which each occurrence affects the probability of further occurrences. These models include the Neyman type A, the Polya-Aeppli, Thomas's Double Poisson, and the negative binomial. An application of three of these models to the problem of the spatial clustering of retail establishments has been described recently by Rogers (1965). Some of the "fits" which he obtained are shown in Figure 3.7.

In the literature of geography, this same problem has received only limited attention. Harvey (1966) has considered the relevance of contagious distributions in the context of spatial diffusion studies. The main contention was that a negative binomial distribution associated with a "randomly distributed colonies" model would be the most appropriate for a two-step-flow diffusion process of the type studied by Hägerstrand (1953). This random colonies model has been discussed by Anscombe (1950, p. 360). "If colonies or groups of individuals are distributed randomly over an area . . . so that the number of colonies observed in samples of fixed area . . . has a Poisson distribution, we obtain a negative binomial distribution for the total count if the numbers of individuals in the colonies are distributed independently in a logarithmic distribution." Harvey suggested that the spatial diffusion of information is basically of the same form and that it is reasonable to assume "that the leaders accept the information at different points in time and that followers accept at a logarithmic rate." Harvey tested his hypothesis using some of Hägerstrand's maps from which systematic samples of quadrats were taken. The frequency distributions of these quadrats, according to the number of points they contained, were compared with the expected distri-

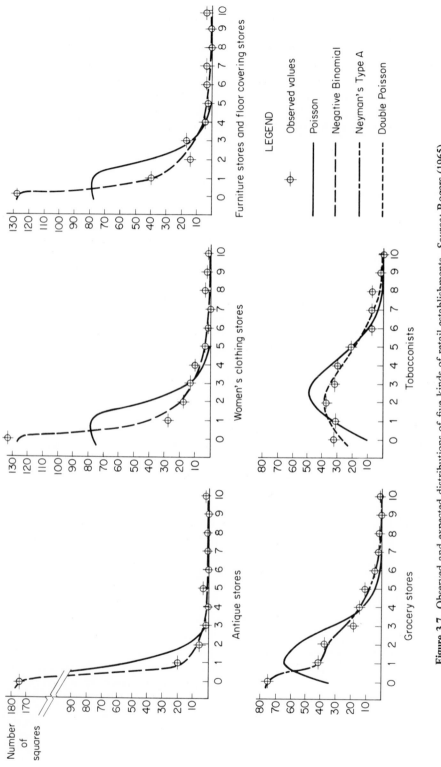

Figure 3.7. Observed and expected distributions of five kinds of retail establishments. *Source:* Rogers (1965).

butions given by several models including the negative binomial. This model is presented in most probability texts, for example, Feller (1950, p. 155) and Parzen (1960, p. 179), as

$$p(x) = \binom{r+x-1}{x} p^r (1-p)^x = \binom{r+x-1}{r-1} p^r (1-p)^x$$

$$= \frac{(r+x-1)!}{x!(r-1)!} p^r (1-p)^x \qquad \text{for } x = 0, 1, \ldots,$$

$$= 0 \qquad \qquad \text{otherwise.}$$

Here r is assumed to be a positive integer and p is such that $0 \leqslant p \leqslant 1$.

The two parameters of the negative binomial law are not derived easily. Williamson and Bretherton (1963) discuss different methods of estimating them. The simplest approach is to use the sample mean m and variance s^2 in deriving p and r as follows:

$$p = m/s^2$$
$$r = mp/(1-p).$$

In the case when r is not an integer, however, then some workers, including Anscombe (1950), prefer to use alternative estimators of the parameters.

Harvey (1966) uses the estimators suggested by Anscombe. His results are illustrated in Table 3.2.

In a subsequent paper, Harvey (1967) has pursued some theoretical considerations concerning the use of the Neyman type A and negative binomial laws in the analysis of diffusion. There are two major issues. One focuses on the effects introduced into the analysis by varying the size of the quadrats. The immediate effect is that the parameters are altered, and Harvey stresses the attendant difficulties of geographic interpretation. The second issue stems from the fact that both of the probability laws under consideration can be derived either as *contagious* models involving generalized Poisson processes or as *compound* Poisson processes for which no assumptions of contagion need be made. The negative binomial model, for example, can be generated by the random colonies process noted earlier which is contagious or by the compounding of a Poisson process for which the parameter λ is a random variable. We shall note in the following section that Dacey (1965a) does in fact derive a special form of the negative binomial law for a point process by compounding a Poisson function in this way. Harvey correctly identifies the problems of trying to draw inferences as to the spatial processes governing point patterns when the empirical evidence affords no way of discriminating between the two possibilities above.

Dacey (1967) has continued the discussion of some of these issues concerning the application of the negative binomial model in geographic studies of location patterns. On an empirical level, Dacey finds that the negative binomial frequency function fits very closely the observed frequency distri-

TABLE 3.2.   NEGATIVE BINOMIAL DISTRIBUTIONS FITTED TO
QUADRAT SAMPLES

Quadrat samples taken from Hägerstrand's (1953) maps of
Columns (1):   Accepters of pasture improvement grants, 1928–33 (Fig. 19, p. 66).
Columns (2):   Accepters of milking machines, 1944 (Fig. 104, p. 290).
Columns (3):   Model II simulated point pattern (Plate III, p. 97).
Columns (4):   Model II simulated point pattern (Plate III, p. 98).

Numbers of points in quadrat	(1)		(2)		(3)		(4)	
	Observed	Theory	Observed	Theory	Observed	Theory	Observed	Theory
0	27	27.1	32	32.2	45	44.8	32	32.3
1	15	13.4	15	16.0	13	14.1	17	14.7
2	9	9.3	9	10.7	9	7.6	10	9.2
3	4	7.0	7	7.6	3	4.7	5	6.3
4	3	5.5	8	5.5	3	3.2	3	4.4
5	4	4.4	5	4.1	3	2.0	4	3.2
6	4	3.6	2	3.1	2	1.4	2	2.4
7	2	3.0	3	2.3	1	1.0	3	1.9
8	3	2.5	1	1.8			0	1.4
9	3	2.1	1	1.4				
10	3	1.8	2	1.1				
11	2	1.5	1	0.8	2	2.2	5.2	5.2
12	4	1.4	1	0.6				
13	1	1.1	3	2.8				
14+	6	6.3						
Chi-square		6.67		2.11		1.10		0.84
Degrees of freedom		6		5		3		4
Level of p		>0.30		>0.80		>0.70		>0.90

Source: Harvey (1966, p. 92).

butions of houses per quadrat for some 21 sample map areas of Puerto Rico.
The fits of this same function also are very good when the quadrats in these
map areas are aggregated into larger units. Against the background of these
empirical findings, Dacey has reviewed, as did Harvey, the different theoretical
models which are known to generate a negative binomial frequency function.
Given this type of methodological situation with a set of empirical findings
on the one hand and a number of possible models on the other, Dacey notes
that a tempting "fail-safe approach to theory construction and verification
is available since any model that generates a negative binomial frequency
function has good correspondence with the observed data," and "because
location theory is sufficiently vague, it is not difficult to construct an intuitively appealing argument that relates settlement distribution to the negative
binomial model."

In rejecting this type of superficial reasoning and analysis, Dacey
stresses the following points:

(i)   The quadrat analysis ignores the arrangement of points in the map space, and when the distances between neighboring points are analyzed, for example, the negative binomial model proves to be inappropriate.

(ii)   The negative binomial model contains the assumption of statistical independence in the sense that the number of houses in any one quadrat is assumed to be independent of the numbers in other quadrats. Now, if the frequency distribution of houses per quadrat is described by a negative binomial frequency function for a given quadrat size, then this assumption of independence demands that as quadrats of identical size are aggregated, the parameter r in the negative binomial model should increase in value in proportion to the increase in area of the quadrats while the other parameter p remains unchanged. This proved not to be the case for the Puerto Rico data, and Dacey concludes that "the negative binomial does not describe the quadrat count data and does not provide an acceptable basis for a theory of settlement distribution."

(iii)   Bias may occur in the description of a location pattern if the parameters of a frequency function are estimated from one size of quadrat only. In general, as the size of quadrat is increased, the degree of clustering observed in the location pattern also increases.

(iv)   The fact that several regularities were observed in the different estimates of the parameters for various-sized quadrats even though the assumption of statistical independence did not hold, suggested the need for a "correlated multivariate negative binomial frequency function." This would give the joint probability of i houses in one quadrat, j in another, k in a third quadrat, and so on when there was a correlation between these numbers of houses for adjacent quadrats. Such a probability function still has to be derived.

Modified and Compound Poisson Laws for Spatial Point Processes

In other work, Dacey has derived a number of appropriate probability laws for particular two-dimensional point patterns. His most important contributions involving discrete probability functions are reviewed here.

The first case Dacey (1964b) considers is that of a point pattern which is more regular than random in a homogeneous region. Homogeneity implies that the density of points throughout the region is constant. The model is given two physical interpretations: one in terms of an urn model, the other as a spatial analogue. We consider here only the spatial interpretation.

The large homogeneous region containing N urban places is partitioned into c equal-sized counties. A distinction is made between urban places (points) which are county seats and those which are noncounty seats. There are Z of the former and $N - Z$ of the latter. There is for each county an equal probability p that it receives a county seat, but no county can receive more than one county seat. This assignment process, Dacey notes, is analogous to sampling without replacement from an urn. The noncounty seats are randomly distributed and each county has an equal probability m of receiving such an urban place. This probability m equals $(N - Z)/c$. In this case a county may receive none, one, or more than one urban place, which is analogous to sampling with replacement. The question is, What is the probability that a county has a total of x urban places? In deriving this probability, two possibilities have to be considered.

(i) If the county has a county seat, then for it to have a total of x places it must have $(x - 1)$ noncounty seat places. The probability of the first event is p; that of the second is given by the Poisson distribution as

$$m^{x-1}e^{-m}/(x - 1)!$$

Hence,

$$P(x') = pm^{x-1}e^{-m}/(x - 1)!$$

where the x prime is used to represent this first possibility.

(ii) If the county does not have a county seat, for which the probability is $(1 - p)$, and it has x noncounty seat places, then

$$P(x'') = (1 - p)m^x e^{-m}/x!$$

The sum of these two mutually exclusive events gives the probability which is sought.

$$P(x) = (1 - p)m^x e^{-m}/x! + pm^{x-1}e^{-m}/(x - 1)! \qquad \text{for } x = 0, 1, \ldots,$$
$$= 0 \qquad \text{otherwise.}$$

Dacey derives the moment-generating function for this law and shows that the mean is $(m + p)$ and the variance is $(m + p - p^2)$. As estimates of the two parameters p and m, the following are given:

$$\hat{p} = [\bar{x} - s^2]^{1/2},$$

and

$$\hat{m} = \bar{x} - \hat{p},$$

where $\bar{x}$ and s^2 are the mean and variance, respectively, for the observed data on points per quadrat. Some empirical fits of this probability function which are taken from Dacey's study are presented in Table 3.3.

In the model above the parameter p is interpreted by Dacey as "a measure of bias toward evenness in an otherwise random arrangement of points" (1966a, p. 174). One extreme is when $p = 0$ and the point pattern is purely random and is described by a Poisson distribution with mean m; for the other extreme, namely $p = 1$, the pattern is more dispersed, with every county receiving at least one point. Dacey notes that in the latter case the probability law $P(x)$ is simply a Poisson variate plus 1.

TABLE 3.3. OBSERVED and CALCULATED FREQUENCY DISTRIBUTIONS
OF URBAN PLACES PER COUNTY IN IOWA, 1840–1960

x	1840 Ob	1840 Ex	1870 Ob	1870 Ex	1900 Ob	1900 Ex	1930 Ob	1930 Ex	1960 Ob	1960 Ex
0	98	98	80	78	35	38	30	29	20	13
1	1	1	16	20	58	53	57	59	60	69
2			3	1	6	7	12	10	15	15
3					0	1	0	1	3	2
4									0	0
5									1	0

Source: Dacey (1964b, p. 565).

The geographic interpretation of the parameters according to Dacey (1966a, p. 175) is that m "represents the mean number of noncounty seat places in each county" and p is the "mean number of county seat places in each county."

In other studies, Dacey has considered more complex situations. First, in his paper (1965a), he discusses an inhomogeneous random point process; that is to say, a Poisson process in which the mean density of points per unit area, defined as λ, is assumed to vary throughout the large region. Specifically, λ is treated as a continuous random variable described by a gamma probability law with parameter γ. Then the probability that a region of area A contains exactly x points is given as

$$P(x; A, \gamma) = A^x(\gamma + x)!/(1 + A)^{\gamma+x+1}x!\gamma! \qquad \text{for } x = 0, 1, \ldots .$$

Dacey identifies this as a special form of a negative binomial distribution, called the *Polya-Eggenberger distribution*. It is not derived in this case, however, as a contagious model. We shall have more to say about this particular study by Dacey when we consider the question of analyzing the distances between neighboring points in a spatial pattern.

Dacey (1966a) also has derived a compound probability law for the county seat–noncounty seat model (discussed first in this subsection) assuming that both of the parameters p and m vary among the counties. Such would be the case in a nonhomogeneous large region. The assumptions made concerning the two parameters again involve continuous probability laws, which we have still to consider. Hence, Dacey's results are summarized only here. The parameter m is assumed to have probability density function $f(m)$, which is a gamma function with parameters γ and β. The corresponding function $f(p)$ for the second parameter assumes that p is a beta variate with parameters r and s. Then the probability $\pi(x)$ that a county contains x places is

$$\frac{r}{r+s} \cdot \frac{(\gamma+x-1)!}{x!(\gamma-1)!}\left(\frac{1}{\beta}\right)^{\gamma}\left(\frac{\beta}{\beta+1}\right)^{\gamma+x}$$

$$+ \frac{s}{r+s} \cdot \frac{(\gamma+x-2)!}{(x-1)!(\gamma-1)!}\left(\frac{1}{\beta}\right)^{\gamma}\left(\frac{\beta}{\beta+1}\right)^{\gamma+x-1},$$

which also may be stated as

$$Rf(x; k, v) + Sf(x-1; k, v),$$

where $f(x; k, v)$ is a negative binomial distribution with parameters k and v. The model is handled as a three-parameter distribution defined by k, v, and S, with $R = (1 - S)$. Estimating equations for these three parameters are presented as part of Dacey's discussion.

Dacey stresses that the probability law $\pi(x)$ identifies only a special case of inhomogeneity and that other forms may prove more appropriate in different situations. Much work still must be accomplished on this question, but Dacey's investigation represents a highly important beginning.

The Normal Probability Law

Attention is focused now on some important continuous probability functions. The first of these is the normal probability law, one of the important foundations upon which the methods of classical statistical inference are built.

The probability density function for this law is given as follows:

$$f(x) = \frac{1}{\sigma\sqrt{2\pi}} \cdot \exp\left[-\tfrac{1}{2}(x - m)^2/\sigma^2\right], \qquad -\infty < x < \infty.$$

The convention is adopted at this point of using the symbol "exp" when the number e is modified by a complicated exponent.

The mean of this distribution is m and the variance is σ^2. The estimators of these parameters are the sample mean $\bar{X}$ and the adjusted variance $s^2/(N - 1)$, respectively.

The graph of this density function is the familiar bell-shaped symmetrical curve which was illustrated in Figure 3.3.

It is worth noting some of the theoretical properties of the normal distribution which make it so important in statistical analysis.

First, for any random variable X which is normally distributed with mean m and variance σ^2, the function $Z = (X - m)/\sigma$ also is normally distributed with mean *zero* and variance *one*. This is a theoretical rationale for the standardization of variables which is mentioned later.

Second, we shall see later that when questions of sampling from a distribution are involved, it is of interest to know what the probability distribution is for any particular statistic computed from the sample. The normal distribution is very important in this context, for it can be shown that if $(x_1, x_2, \ldots, x_n)$ is a random sample from a normal distribution with mean m and variance σ^2, then the sample mean $\bar{X}$ also is normally distributed with mean m and variance σ^2/n. It follows that the random variable $\dfrac{\bar{X} - m}{\sigma/\sqrt{n}}$ is normally distributed with mean *zero* and variance *one*.

Third, there is an important theorem, the *central limit theorem*, which states that, for *any* distribution with mean m and a finite σ^2, as the sample size n increases, the distribution of $\bar{X}$ approaches a normal distribution with mean m and variance σ^2/n.

We shall see later that these properties are very important in the context of statistical inference.

Curry (1964) has shown that for a randomly located set of settlements the distances between the settlements and their nearest neighbors will be normally distributed. The derivation in this case is with respect to a geometric representation of the two settlements as points in a Cartesian space. The probability of a settlement being located within a distance range of u and $(u + du)$ units from its nearest neighbor is given as the normal distribution. This suggestion that the distances to nearest neighbors in a random pattern are normally distributed is implicit in the studies by Getis and King mentioned earlier.

The Lognormal Probability Law

Another useful property of the normal distribution is that many other distributions can be changed into a form approaching normal by the use of transformations. For example, a random variable X may not be normally distributed, but the distribution of the variable $Y = \ln X$ may be normal. In this case, X is said to be lognormally distributed.

The lognormal distribution is but one of a number of highly skewed distributions which have been discussed extensively in the literature of social science, particularly economics. Many of the writings in the latter field were concerned with the distributions of income or business firm sizes. The more important of these are reviewed in a recent book by Steindl (1965).

In discussing this law, Aitchison and Brown (1957) and earlier workers such as Kapteyn (1903) and Gibrat (1931) suggested that it would be generated by the "law of proportionate effect" which can be stated as follows: "In a process of growth, equal proportionate increments have the same chance of occurring in a given time-interval whatever size happens to have been reached. In other words, growth in proportion to size is a random variable with a given distribution which is considered constant in time" (Steindl, 1965, p. 30).

In their different studies of the location pattern of urban settlements, Thomas (1962), Curry (1964), and King (1961) have discussed the lognormality of the distances between towns and their nearest neighbors of the same size.

The distribution of city sizes in many countries also has been shown to be lognormal (Figure 3.8). In discussing this situation, Berry (1961a) has argued that "as a limiting case, a lognormal distribution is a condition of entropy, defined as a circumstance in which the forces affecting the distribution are many and act randomly," and hence lognormal rank-size distributions "are found when, because of complexity of economic and political life and/or age of the system of cities many forces affect the urban pattern in many ways." Thomas (1967) has outlined a more formal derivation of a lognormal distribution of city population size.

Another skewed distribution which has been used to describe the distribution of city sizes is the Pareto law. The density function for this law is given by Parzen (1960, p. 211) as

$$f(x) = rA^r(1/x^{r+1}) \qquad \text{for } x \geqslant A,$$
$$= 0 \qquad\qquad \text{for } x < A,$$

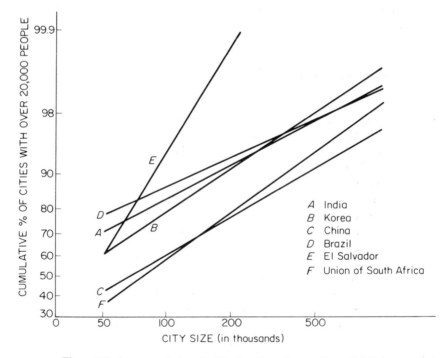

Figure 3.8. Lognormal city size distributions. *Source:* Berry (1961a).

where r and A are positive parameters. Quandt (1964) has noted other types of distributions which are identified also as Pareto distributions.

Singer (1936) and Allen (1954) discussed the application of the Pareto law to city sizes, and Morrill and Pitts (1967) used the function in their analysis of individual information fields.

Finally, with respect to the lognormal and Pareto distributions, the work of Kulldorff (1955) should be noted. In studying the probabilities that migrations beginning in a region will cross the border and terminate outside the region, Kulldorff makes certain assumptions concerning the distribution of migration distances. Among these are the suggestions that the distances are distributed first according to the Pareto law and second, according to the lognormal distribution. On the bases of these assumptions and others relating to the distribution of starting points and the direction of movements, the migration probabilities were derived for different-shaped regions.

The Exponential Probability Law

This is another important continuous probability law. In its most common form, the model is written as

$$f(x) = ce^{-cx} \qquad \text{for } x \geqslant 0,$$
$$= 0 \qquad \text{otherwise.}$$

A variation on this model has been used in urban geography to describe the decline in population density over distance from the city center. Berry, et al. (1963), for example, discuss a model of the form

$$d_x = d_0 e^{-bx},$$

where d_x is the density at a distance x from the city center and d_0 is the density extrapolated for the center. It should be stressed that in this form the function is not a probability density function and cannot be integrated to give a value of 1. The authors did not treat this model as a formal probability distribution, however, and it is referred to here only as an example of a negative exponential function. Morrill and Pitts (1967) similarly fit exponential functions to migration and marriage distances data.

The Gamma Probability Law

This is a two-parameter probability distribution, which is a general one in the sense that some other distributions such as the lognormal can be derived as special cases of it. The density function is given by Parzen (1960, p. 180) as

$$f(x) = \frac{\lambda}{(r-1)!}(\lambda x)^{r-1}e^{-\lambda x} \qquad \text{for } x \geqslant 0,$$

$$= 0 \qquad \qquad \text{otherwise.}$$

The parameters are defined as $(r = 1, 2, \ldots)$, and $(\lambda \geqslant 0)$.

Dacey's use of the gamma function in deriving a compound distribution for a point process more uniform than random with areal inhomogeneity has been referred to earlier. For the density parameter m the probability was given by the function

$$f(m) = m^{\gamma-1}e^{-m/\beta}/\beta^\gamma(\gamma-1)! \qquad \text{(Dacey 1966a, p. 175)}.$$

This expression is obtained from that given by Parzen by substituting $x = m$, $r = \gamma$, and $\lambda = 1/\beta$.

Dacey (1963) has shown that the gamma distribution also gives the probability for the distance from an arbitrary locus to the jth nearest point in a two-dimensional homogeneous Poisson point process. The derivation is not a simple one and no attempt is made to present it here. In the case of inhomogeneous random point patterns of the type described by the Polya-Eggenberger distribution, the probabilities for these same distances are given by another density function which Dacey (1965a) has derived.

A geographic-type application of the gamma probability model has been outlined by Decker (1952) in a study of the probability of occurrence of damaging hail storms over a 20-year period in Iowa. For four regions of the state, the probabilities of experiencing x dollars of hail damage per 1000 acres were evaluated by the fitting of gamma distributions to the regional data.

The particular functions discussed in this section do not exhaust all the known continuous probability laws and the reader is referred to a book such as Parzen (1960, pp. 176–182) for a review of other important ones. The functions mentioned above, however, are those which have appeared in the geographic literature. Also, as Dacey's work illustrates, the problem under consideration often demands the derivation of compound distributions and there are many possibilities along these lines which still have to be introduced into the geography literature. Nor has there been any serious progress made in obtaining the joint probability laws for spatial problems involving several random variables (Parzen 1960, pp. 329–334).

3.3 TWO-DIMENSIONAL PROBABILITY SURFACES

By virtue of his interest in locational analysis, the geographer typically deals with observations which are two-dimensional in the sense that they can be defined with respect to a pair coordinate axes. The latitude-longitude grid is one such pair. In this section, however, we consider the axes x and y of a rectangular or Cartesian space and we shall note briefly some probability functions which can be defined in these two dimensions. Unfortunately this topic has received little formal attention in the geographic literature, although it is implicit in some studies.

In an unpublished review of the general problem of computing population potentials, Court (1966) has discussed the application of selected two-dimensional models. He notes that there are particular mathematical distinctions about the form of these models that must be clarified.

The expression $p(x, y)$ is a density function for the points having coordinates x and y in a Cartesian space. This probability also can be given in terms of polar coordinates. The distinction between rectangular and polar coordinates is illustrated in Figure 3.9. In polar coordinates the probability referred to above is $p(v, \theta)$. The mathematical transformation from rectangular to polar coordinates is not difficult, but Court notes that there may be attendant problems in interpretation of the different functions for a phenomenon such as population density. Whereas in rectangular coordinates the density for a small unit square is everywhere the same, in polar coordinates the density per unit sector segment increases away from the origin.

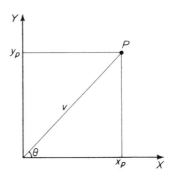

Figure 3.9. Polar and rectangular coordinates. The point P in the plane is uniquely determined by its rectangular coordinates (x_p, y_p) and its polar coordinates (v, θ).

Another distinction which must be made is between a *conditional* and *marginal* radial density. Assuming that population density is to be described in terms of radial distance, then Court expresses the distinction as follows: ". . . for a given value of θ, say θ_0, the conditional density of v can be obtained as $p(v \mid \theta_0)$. It may be visualized as the profile, along the ray $\theta = \theta_0$, of the bivariate distribution represented by $p(v, \theta)$." If this conditional density is the same for all θ, the distribution is said to be *circularly symmetric* or *isotropic*.

The marginal radial density "is the sum of all the conditonal densities. It is obtained by integrating around the circle . . . *and* . . . may be visualized as the result of sweeping a board around the bivariate distribution, piling up along one radius all the population that originally was distributed over the plane."

In reviewing some of the applications of radial probability distributions in the literature of the social sciences, Court is critical of the repeated failures to differentiate between the conditional and marginal densities, rectangular and polar coordinates, and functions which are handled either as probability laws or statistical descriptions of situations.

In some of the recent work by Russian geographers there are attempts to develop probability models of two-dimensional distributions. The studies of urban population densities by Gurevich and Saushkin (1966) and Gurevich (1967) are noteworthy in this context. Dacey (1965f) has discussed the use of the circular normal distribution in describing the distribution of population around a city center. This line of work undoubtedly will be developed further in the future.

In this section so far we have been dealing with *radial* distributions. In many sciences, for example, geology and biology, *circular* distributions are employed quite frequently. In this case the total probability is spread out on the circumference of a circle. These circular distributions are appropriate for observations which relate to directions in a plane and which can be "represented by dots on the circumference of a unit circle" (Batschelet, 1965, p. 11). In geology, circular normal distributions are postulated for the orientation of rock joints and fractures and for particle axes in sediments (Krumbein and Graybill, 1965, p. 110).

3.4   THE FITTING OF PROBABILITY MODELS

Throughout this chapter there have been many references made to applications of the different probability distributions in geographic research. A probability distribution is a model and as such it may be fitted to any set of empirical observations. The application of probability models in applied research usually demands that there is some way of giving particular numerical values to the parameters involved in the model. There are, in fact, at least two important questions raised here. The first concerns the actual method of providing such numerical values for the parameters, and the second has to do with deciding on how well the model actually fits reality.

The estimation of parameter values is an issue in statistical inference and as such it is discussed in detail in the following chapter. There are two methods used most frequently in estimating parameters. One is the *method of moments*, in which the sample moments are used as the basis for estimating the model parameters. The estimators which Dacey derived for the parameters in his modified Poisson distribution exemplify this approach. A more formal probability approach is by way of deriving the *maximum likelihood* estimators. Consideration is given to these topics and testing the goodness of fit in the next chapter.

A more basic question is repeated at this point. It is possible that more than one probability model will fit a set of empirical observations equally well. The problem then arises of choosing among the models. Hopefully, the broader theoretical context within which the statistical analysis is set would suggest which model is more appropriate. Indeed, it would seem desirable that the theory, or at least the hypotheses from which the study is proceeding, should suggest by way of logical reasoning or deduction what form the probability distribution should take. This model conceptualization, then, would precede any fitting or empirical testing of the probability model. But research procedures in geography to date have seldom been so formal, and there exist a number of studies in which curve fitting has been pursued as an end in itself with little concern given to the possible correspondence rules between the model constructs and observable realities.

A point can be made in partial defense of the curve fitting, however. To the extent that computer simulations are being attempted for many dynamic phenomena, it is often necessary to have available some summary characteristics of factors or variables which are to serve as inputs to the simulation model. This is especially so for a type of simulation procedure known as the Monte Carlo method. Given these demands, it may be necessary to describe many features of reality by way of probability models for which correspondence rules with reality are only weakly structured, if at all.

SUGGESTED READINGS

Berry, Brian, J. L., and Garrison, William L. (1958), "Alternate Explanations of Urban Rank-Size Relationships." *Annals*, Association of American Geographers, Vol. 48, pp. 83–91.

Clark, P. J., and Evans, F. C. (1954), "Distance to Nearest Neighbor as a Measure of Spatial Relationships in Populations." *Ecology*, Vol. 35, pp. 445–453.

Curry, L. (1964), "The Random Spatial Economy: An Exploration in Settlement Theory." *Annals*, Association of American Geographers, Vol. 54, pp. 138–146.

Dacey, M. F. (1964a), "Two-dimensional Random Point Patterns: A Review and an Interpretation." *Papers*, The Regional Science Association, Vol. 13, pp. 41–55.

Dacey, M. F. (1964b), "Modified Poisson Probability Law for Point Pattern More Regular Than Random." *Annals*, Association of American Geographers, Vol. 54, pp. 559–565.

Fraser, D. A. S. (1958), *Statistics: An Introduction*. New York: John Wiley & Sons, Inc., Chapters 1–5.

Hoel, P. G. (1960), *Elementary Statistics*. Second edition. New York: John Wiley & Sons, Inc., Chapters 3–4.

Kemeny, J. G., Snell, J. L., and Thompson, G. L. (1966), *Introduction to Finite Mathematics*, Second edition. Englewood Cliffs, N.J.: Prentice-Hall, Inc., Chapters 2–4.

Rogers, A. (1965), "A Stochastic Analysis of the Spatial Clustering of Retail Establishments." *Journal of American Statistical Association*, Vol. 60, pp. 1094–1103.

ROLE OF STATISTICAL
INFERENCE IN GEOGRAPHY

4

Statistical inference is concerned with the estimation of parameters and the testing of a broad range of statistical hypotheses. The approach involves making use of information associated with a *sample* drawn from the *population* or distribution of interest. For this reason, statistical inference and sampling theory are very closely interrelated.

This chapter deals with some basic problems in statistical inference and sampling as they have appeared in the literature of geography. The examples are few, for by the very nature of his interests the geographer deals only infrequently with samples. More often he is concerned with the total set of observations, and inferential statements are not always meaningful in this context. Also, some of the sampling procedures employed by the geographer, for example, traverses, do not allow the use of the formal theory that is possible with other sampling designs. Some of these latter designs are reviewed briefy in the first section of this chapter.

4.1 SAMPLING DESIGNS FOR SPATIAL PATTERNS

Most discussions of sampling distinguish between the *universe* or *population* as the complete set of observations or cases and the *sample* which is a subset of the universe selected for the purpose of representing that same universe. This distinction is not always adhered to in research, however. For example, Thomas and Anderson (1965), in their study of certain technical problems in spatial correlation analysis, found it convenient to distinguish between the universe and population. The former was defined as "a more abstract group than population and contains all events as they happened and as they might have happened if everything else had remained the same but the random shocks."

We might note also the distinction which often is drawn between the *target* population and the *sampled* population. Cochran, Mosteller, and Tukey (1954) discuss this distinction, and Krumbein and Graybill (1965, pp. 149–153) mention it with regard to geological sampling. They note that there is often "a conceptual population based on the objectives of . . . geological study . . . whose parameters it is desired to determine," but for which "most of the individuals are not available for sampling." It is this *target* population at which the study is aimed, but the sampling has to be done from the *sampled* population which is available for study. There appear to be many opportunities for emphasizing this distinction in geographic research, but unfortunately it has not been pursued. In land-use studies, for example, the target population might be conceptualized as the kaleidoscope of patterns that exists from month to month and year to year. The sampled population is the pattern which exists at the particular time a survey is taken.

A statistical sampling procedure requires that the members of a sampled population can be identified and in many cases enumerated and that the probability of any member being selected can be stated. Most geographic research involves clearly defined populations which may or may not be finite in size. The population of iron and steel plants in the United States is a finite set, whereas the number of points in a map area is infinite.

The sampling unit in most geographic studies is either a *point* or a *quadrat*. This is not to overlook the fact that many geographic studies are concerned with the sampling of individual households, persons, manufacturing plants, and so on, but these sampling units are not explicitly spatial and are common to the work of other social scientists. As such, they are discussed already in a number of good texts such as Cochran (1953) and Kish (1965). The second of these books contains also an extended treatment of the problems of area sampling (Kish, 1965, Chapter 9), and the reader is referred to this discussion. The following review emphasizes point sampling techniques.

An essential requirement in statistical sampling is that the selection of the sample units be made in such a way as to ensure that the units chosen are representative of the population. Every effort is made to avoid any *bias* associated with giving undue emphasis to particular members of the population. Some of the commonly employed sampling procedures which satisfy these requirements are discussed below.

Random Sampling of Spatial Patterns

Random sampling is implied in most theoretical discussions of statistical inference. The method ensures that the members of a population have equal probabilities of being selected and that the selection of a unit in no way influences the selection of any other one. The sample members are drawn independently with equal probabilities.

By using a grid overlay on a map, a population of coordinate points can be generated. Every one of these points can be identified by its X and Y coordinates, and the use of a random numbers table allows for the selection of a random sample of these points. The use of this table is described in most books of statistical tables. Usually, a starting point within the table is chosen randomly; then, in accordance with some predetermined way of working through the table, numbers of an appropriate digit size are selected. Whenever a selected number corresponds to the coordinates of a point in the area, then that point is included in the sample. Repeated occurrences of the same number generally are ignored, although they need not be, and the selection continues until the sample is as large as required.

Again with respect to any such map the quadrats associated with the grid could be numbered consecutively and a random sample of these chosen in the same way as with the points. Alternatively, the quadrats might be identified by the coordinates of their center points. These designs might be appropriate in land-use surveys.

An advantage of working with random samples is that an important assumption underlying many statistical techniques is satisfied as a result. The sample observations are independent and have equal probabilities of being chosen. On the more practical side, however, there is the disadvantage that the procedure does not always ensure an adequate areal coverage of a study area. Some clustering of the sample points often results and, as a consequence, important spatial properties may be overlooked.

Systematic Point Sampling

To avoid the shortcomings of random sampling regarding the coverage of an area, a systematic sampling plan may be used. This procedure demands that it be possible to work through the set of population members in some consistent and orderly manner, such as along a row of quadrats or a line of coordinate points. The sampling is initiated by the random selection of a point or quadrat; then in accordance with a predetermined plan the remainder of the sample is selected.

Quenouille (1949) and Berry (1962) have discussed the problem of systematic sampling from sets of coordinate points. Two basic designs are considered. One has points *aligned* in a checkerboard fashion; in the other the points are *unaligned*. For the first of these designs, the study area is divided into a set of grid squares all containing the same number of coordinate points. One point is chosen randomly in the first cell; the *corresponding* point locations in the other cells are selected then as the remaining sample members (Figure 4.1).

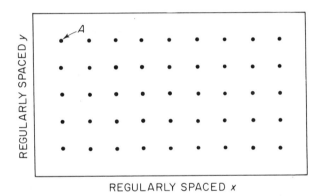

Figure 4.1. A systematic aligned (checkerboard) sample. A checkerboard sample has a perfectly even spread of points, with regular spacing on both abscissa and ordinate after point *A* has been located at random. But such a selection procedure implies that all parts of the study area do not have an equal chance of being included in the sample. Furthermore, if there are periodicities in the data being collected, the regularly spaced points could hit the same point on a cycle time and again, and give completely biased pictures of the spatial variations of phenomena under study. *Source:* Berry (1962).

With the unaligned point design, the first point is chosen in the same way as for the aligned case. The *X* coordinate of this point in the first cell is then held constant for all the other grid squares across the top row and a point in each of these squares is chosen by randomly selecting new *Y* coordinates. Similarly, for the other cells in the first column, the *Y* coordinate of the point in the first cell now is held constant while the *X* coordinate is randomized.

In Figure 4.2, this unaligned procedure yields points *B*, *E*, and so on

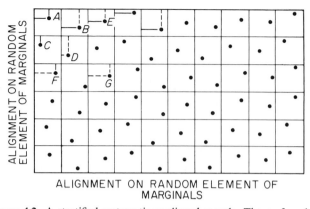

ALIGNMENT ON RANDOM ELEMENT OF
MARGINALS

Figure 4.2. A stratified systematic unaligned sample. The preferred areal sample is the stratified systematic unaligned sample. It is constructed as follows: First, point *A* is selected at random. The *x* coordinate of *A* is then used with a new random *y* coordinate to locate *B*, a second random *y* coordinate to locate *E*, and so on across the top row of strata. By a similar process the *y* coordinate of *A* is used in combination with random *x* coordinates to locate point *C* and all successive points in the first column of strata. The random *x* coordinate of *C* and *y* coordinate of *B* are then used to locate *D*, of *E* and *F* to locate *G*, and so on until all strata have sample elements. *Source:* Berry (1962).

in the top row, and *C*, and *F*, and so on in the first column. The selection continues as follows: "The random *X* coordinate of *C* and *Y* coordinate of *B* are then used to locate *D*, of *E* and *F* to locate *G* and so on until all strata have sample elements" (Berry 1962, p. 7).

Berry (1962, pp. 8–11) notes that, on the basis of empirical studies of land-use sampling, the relative efficiency of systematic unaligned designs is relatively high. Efficiency is defined in terms of the ratio of the variance associated with some other design to the variance of the systematic unaligned sample. The higher the ratio, the greater the relative efficiency of the systematic unaligned design.

It should be noted that for the aligned design the selection of the first point determines the selection of the remaining sample members. Therefore, it is not the case that all points have an equal probability of being chosen once the first member of the sample is determined. In the context of estimation problems, this gives rise to some important mathematical questions. The reader is referred to Quenouille (1949), Yates (1953), or Williams (1956) for a discussion of how some of the attendant difficulties can be resolved.

Finally, it is worth stressing with regard to the systematic aligned sample that "if there are periodicities in the data being collected, the regularly spaced points could hit the same point on a cycle time and again, and give completely biased pictures of the spatial variations of phenomena under study"

(Berry 1962, p. 7). This problem is alleviated by the use of the systematic unaligned design.

Stratified Areal Sampling

In the preceding discussion of systematic point sampling, it was assumed that the area was divided into a number of squares, each containing a set of points. This amounted to a *stratification* of the area, a strategy which often is adopted in sampling problems to take into account existing knowledge of the population under consideration. For instance, in sampling to ascertain the number of first-generation immigrant families within a large metropolitan area, it would be relevant first to zone the city according to the distance from downtown and then to sample within each zone. Since the density of

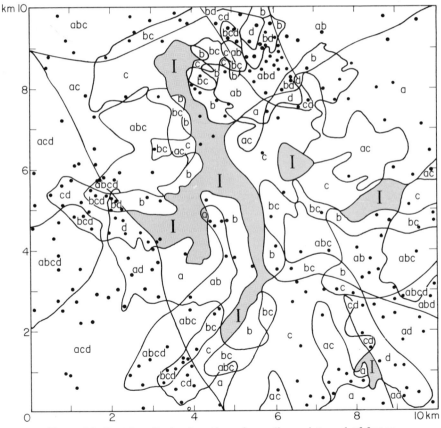

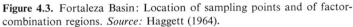

Figure 4.3. Fortaleza Basin: Location of sampling points and of factor-combination regions. *Source:* Haggett (1964).

population falls off with distance from downtown, the sample size in each zone could be adjusted accordingly.

Haggett (1964) has discussed the use of a stratified sample in a study of the distribution of forest in the Fortaleza basin of eastern São Paulo. The basin first was stratified with respect to four possible explanatory factors and then ". . . for each of these factor combination types sixteen sampling plots were located on the map using random coordinate methods . . . ." The resultant sample of points is shown in Figure 4.3. These points then were identified on aerial photographs as circles of 2.5 acres in area, and the proportions of these areas covered in forest were calculated.

Wood (1955) outlined the application of a stratified random design in a land-use study of the swamps and marshes of eastern Wisconsin. In this study, there was multistage stratification. First, the universe of 290 "towns" was ranked according to the amount of wet land they had at the time of an earlier land survey. Five groups were identified from this ranking, and then each group was divided further into "much decrease" and "little decrease" subgroups. "The plan was to draw one town from each of the ten subgroups to form the sample." The subsequent random selection of a town in each group was constrained, however, by the requirement that each of three larger regions, the south, central, and north, had to be represented in the sample by at least three but no more than four of the towns.

Hierarchical Sampling

In some studies it may be desirable to sample with increasing detail at a number of levels. Thus, we might first select randomly one or more counties in a state. Then within these counties we might sample a number of quadrats or, say, townships and finally, within the latter, randomly select some farmsteads. This was the type of strategy followed by Wood in the study cited above.

A related sampling design, "cluster sampling" is employed often in social surveys, and Kish (1965, Chapters 5, 6, 10) and Sampford (1962) provide excellent discussions of the technique. In deriving estimates from such samples, the variance contributed at every level of sampling can be determined.

Traverses

The use of cross-sectional lines or traverses has always been favored in geographic research, although the reliability of the estimates derived in this way has never been examined rigorously. One interesting study, however, has been completed by Haggett and Board (1964). They examined the comparative efficiencies of parallel and rotational traverses in estimating geographic areas and concluded that in situations where orientation was not important the parallel traverses yielded more efficient results. The use of traverses in other sciences, particularly botany and forestry, is well established (Cain and Castro, 1959).

Other Approaches in Spatial Sampling

Mention was made in the preceding paragraph of forestry as a science concerned with problems of sampling spatial distributions, and this interest is shared by a number of the agricultural sciences. Snedecor (1956) and Greig-Smith (1964), for example, include sections on sampling in their books. Many of the sampling designs used in these other sciences are explicitly spatial, and it behooves the geographer to be aware of them. Matern (1960), for example, deals with a number of sampling designs which appear relevant in the context of geographic research. His concern is over methods for locating sample points in a two-dimensional region R_2, assuming that the basic strategy involves a subregion Q in R_2 and the sampling of a finite number of points in Q. The sampling designs include simple random procedures; stratifications based on nonoverlapping geometric forms such as the circle, hexagon, and square; Latin square designs, in which first a set of n^2 plots is arranged in n rows and n columns with a sample of the plots being drawn in such a way that every column and row is represented and then a point is randomly chosen in each of the sample plots; and, finally, systematic designs for which the sample consists of a lattice of points forming a regular geometric figure such as a triangle, hexagon, or square.

There are a number of other questions related to spatial sampling which command Matern's attention. The cost efficiency of different designs expressed as a function of the distance between sample points is one such question. In addition, "border effects," line surveys as one-dimensional samples of points, the size of sample tracts, and the location of secondary sampling plots on the circumference of these tracts are topics which are discussed in some detail. Most of these same questions are of interest to the geographer, but to date they have not been considered by him.

4.2 SAMPLING DISTRIBUTIONS

If we assume in sampling that all members of the population have an equal chance of being selected, then it is quite conceivable that our sampling might pick up mainly extreme values on one occasion and quite different values on another. Therefore, any *statistic* which is computed from the sample data will differ in value from sample to sample. Assuming an infinitely large number of samples, it is possible to think of a probability function which would give the probabilities of the statistic value falling in certain specified intervals. Such a probability function is known as a *sampling distribution*.

There are two general observations to be made at this point. First, the mathematical properties of these sampling distributions are deduced from the known information concerning the parent distribution. These deductions belong to the realm of mathematical statistics. Second, in any research problem we are concerned usually with only one sample and one value for the statistic, but it is important in hypothesis testing to know the sampling distribution of the statistic involved.

For some statistics the sampling distributions are the probability laws which we have discussed already. This is the case, for example, with regard to the *mean*. As noted, if the sample is from a normally distributed parent population, then the distribution of sample means is also normal. In addition, the *central limit theorem* states that the sampling distribution for the mean will be normal regardless of the character of the parent population, as long as the sample size is large, that is, more than about 30. The normal distribution does not apply in all cases, however, and this section begins with a discussion of three other important sampling distributions: the chi-square, "*t*," and "*F*" distributions. These are continuous probability models and could have been discussed earlier along with the other continuous models. They are employed more frequently in the context of statistical inference, however, and it is convenient to introduce them at this point.

Chi-square Distribution

Consider a standardized normal variate Z; that is, Z is normally distributed with mean zero and variance one. Let $(z_1, z_2, \ldots, z_n)$ be a sample of values from this distribution. Then the statistic formed by the sum $(z_1^2 + z_2^2 + \ldots + z_n^2)$ is defined as χ^2, and this has the probability density function

$$P(\chi^2) = \frac{1}{2^{n/2}\Gamma\frac{n}{2}} \cdot e^{-(\chi^2/2)} \cdot (\chi^2)^{(n-2)/2}, \qquad \text{for } \chi^2 > 0,$$

$$= 0 \qquad\qquad\qquad \text{otherwise.}$$

The symbol Γ denotes a mathematical function, the gamma function, which is defined for any positive integer n as $(n - 1)!$ The function is continuous, however, and can be defined for any positive real number (Fraser, 1958, pp. 190–191). The function is given in Appendix B.

The chi-square distribution has the one parameter n, and as this increases, the distribution curve becomes more symmetrical. The parameter n is also the number of degrees of freedom for the statistic χ^2.

The following properties of the chi-square distribution are important:

(i)   For n greater than about 30, the transformation $Z = (2\chi^2)^{1/2} - (2n - 1)^{1/2}$ allows for the use of the standardized normal distribution.

(ii)   An additivity property. If χ_1^2 and χ_2^2 are independent random variables having chi-square distributions with n_1 and n_2 degrees of freedom, then $\chi^2 = \chi_1^2 + \chi_2^2$ is distributed also as chi-square with $(n_1 + n_2)$ degrees of freedom.

(iii)   It can be shown that the statistic

$$X^2 = \sum_{j=1}^{k} \frac{(f_j - F_j)^2}{F_j},$$

where f_j is the observed frequency in the jth of k classes and F_j is an expected or theoretical frequency for the same class, is distributed approximately as chi-square with $(k - 1)$ degrees of freedom. This statistic is important in testing the goodness of fit of two distributions or the independence of two classifications.

The "t" Distribution

Consider a standardized normal distribution. Assume that one value (z_0) is drawn randomly, and then another n values of the population $(z_1, z_2, \ldots, z_n)$ are sampled. Recall that

$$\chi^2 = z_1^2 + z_2^2 + \ldots + z_n^2.$$

Then the statistic

$$t = z_0/(\chi^2/n)^{1/2}$$

has the probability distribution

$$p(t) = \frac{1}{(\pi n)^{1/2}} \cdot \frac{\Gamma\left(\frac{n+1}{2}\right)}{\Gamma\left(\frac{n}{2}\right)} \cdot \left(1 + \frac{t^2}{n}\right)^{(-n+1)/2} \qquad \text{for } -\infty < t < \infty.$$

Again, the shape of the curve depends upon the number of degrees of freedom n, and as this increases in size, the "t" distribution approaches the normal distribution in form. In contrast to the chi-square, the "t" distribution is symmetrical.

This is a particularly important distribution in statistical inference. For example, if the sample is from a normal population with mean μ and variance σ^2, then the statistic

$$t = \frac{\bar{X} - \mu}{s/\sqrt{n}},$$

where $\bar{X}$ is the sample mean, s the sample standard deviation, and n the sample size, has a "t" distribution with $(n - 1)$ degrees of freedom. This is the basis for certain tests of hypotheses concerning population mean values.

The "F" Distribution

Consider two independent chi-square random variables χ_1^2 and χ_2^2 with n_1 and n_2 degrees of fredom, respectively. Then the statistic

$$F = \frac{\chi_1^2/n_1}{\chi_2^2/n_2}$$

has the probability density function

$$p(F) = \frac{\Gamma\left(\dfrac{n_1 + n_2}{2}\right)}{\Gamma\left(\dfrac{n_1}{2}\right)\Gamma\left(\dfrac{n_2}{2}\right)} \cdot \left(\frac{n_1}{n_2}\right)^{n_1/2} \cdot \frac{F^{(n_1-2)/2}}{\left(1 + \dfrac{n_1 F}{n_2}\right)\left(\dfrac{n_1 + n_2}{2}\right)} \cdot$$

In contrast to the two preceding functions, this is a two-parameter distribution (n_1, n_2), and the published tables for the distribution have to be entered with both of these degrees of freedom.

One important property of the "F" distribution is that for random samples of sizes n_1 and n_2, drawn from normal populations with variances σ_1^2 and σ_2^2, respectively, the ratio of the sample variances s_1^2/s_2^2 has approximately the same distribution as $\dfrac{\sigma_1^2}{\sigma_2^2} \cdot F$, with $(n_1 - 1)$ and $(n_2 - 1)$ degrees of freedom. This property provides the basis for some important tests of significance.

All three of the sampling distributions above have been tabled (Hoel 1960, pp. 243–252; Fraser 1958, pp. 387–394). In the following sections, the use of these tables in problems of statistical inference will be assumed, and the reader should familiarize himself with these tables.

4.3 ESTIMATION OF PARAMETERS

An important topic in statistical inference is estimation of the unknown parameters of populations or distributions. This topic is not well represented in the geography literature, however, and only a few summary comments will be made here on the problem.

Estimation is approached usually in one of two ways, although there are other methods available. The first involves *point estimation*, whereby single values are derived for the parameters of interest. In this approach, there are a number of properties of an estimator which are sought. It should be *unbiased* in the sense that the expected value of the estimator should equal the true value of the parameter in question, *efficient* in that it has the smallest variance of any of the unbiased estimators, and, hopefully, *sufficient* in the sense that it summarizes all the information about the parameter that can be derived from an outcome. Fraser (1958, pp. 214–246) and Hogg and Craig (1959) provide detailed discussions of these properties.

Point estimates are derived in a number of ways. The only approach used in geographic research to date has been the *method of moments*. The different moments for the sample are used as estimates of the corresponding population moments. Dacey, in his work on probability functions for point processes which was reviewed in the previous chapter, uses estimates derived in this way.

A more formal method of point estimation is the *maximum likelihood* approach. Given a particular sample outcome $(x_1, x_2, \ldots, x_n)$, the estimate is derived in such a way that the relative probability or likelihood of the outcome is maximized. The method is discussed in Fraser (1958, pp. 224–228).

The use of linear models of the form $y_i = \alpha + \beta x_i + e_i$, where α and β are parameters, also involves problems of point estimation. The associated method of least squares, which has still to be discussed, is indeed a form of estimation. Problems of estimation in this context are discussed at length in the econometrics literature, for example, Johnston (1963) and Malinvaud (1966).

An alternative approach to estimation is that of constructing *confidence intervals*. In estimating a parameter θ, two values $\underset{\sim}{\theta}$ and $\bar{\theta}$, which are functions of the sample data, are sought such that over the long run the probability that intervals of the form $(\underset{\sim}{\theta} \leqslant \theta \leqslant \bar{\theta})$ will contain θ is given as γ, where γ is a selected probability value or level of confidence. This approach uses the sample statistic corresponding to the parameter of interest and knowledge of the sampling distribution for that statistic. Thus, in the case of a large random sample (at least 30) drawn from a normal population, it can be shown that a 95 percent confidence interval for the population mean μ is given by the expression

$$\left(\bar{X} - \frac{1.96s}{\sqrt{n}}\right) \leqslant \mu \leqslant \left(\bar{X} + \frac{1.96s}{\sqrt{n}}\right),$$

where n is the sample size, s is the sample standard deviation, and 1.96 is the value of "t" with $(n - 1)$ degrees of freedom for the 95 percent probability level.

Confidence intervals can be derived for many other parameters. Assume, for example, that the problem is to estimate the proportion of a total map area which is forested. On the basis of the evidence from a point sample, it is possible to compute a confidence interval for this proportion. The sample statistic in this case would be the proportion of all the sample points at which forest is recorded. The corresponding confidence interval for a given probability level can then be read from the standard charts which are reproduced in most statistical texts, for example, Krumbein and Graybill (1965, p. 441).

Those sampling designs which allow for the deriving of fairly narrow confidence intervals for the parameters in question usually are preferred. To date, geographers have largely ignored the question of confidence intervals for spatial parameters.

4.4 TESTING HYPOTHESES IN A GEOGRAPHIC CONTEXT

The testing of hypotheses is a major area of interest in statistical inference. These hypotheses typically relate to parameter values or to the differences between parameter values. Again, the applications in the geography literature are few, but there are some interesting ones. Therefore, it is useful to review briefly the logic of hypothesis testing as an introduction to these examples.

Consider a simple hypothesis to the effect that the average distance traveled by farm families to purchase groceries in a region is 5 miles. Assume it is known on the basis of past studies that these distances have a variance of 4 miles, in which case the standard deviation is 2 miles. A random sample of 36 farmers is to be selected for the purpose of testing the hypothesis above. We shall use this sample size to facilitate computation.

In a formal statistical sense, the hypothesis which is tested in such situations is that there is no significant difference between the true mean value for the universe (the average distance traveled for all farmers in the region) and the hypothesized value, in this case 5.0. This hypothesis of no difference is the *null hypothesis*, H_0. The alternative hypothesis, H_a, in this example is that the true mean is either greater or less than 5.0.

How is the sample evidence to be used in testing the null hypothesis? First, a test statistic has to be decided upon, and since we are interested in the population mean, it is appropriate to use the sample mean as the test statistic.

Now, a knowledge of sampling error suggests that in drawing our sample we might obtain a set of quite extreme values no matter how unbiased our sampling procedure is and that the sample mean might be "weighted" accordingly. Indeed, we know that there is a whole distribution of possible outcomes for any statistic, namely the sampling distribution. What we have to decide is which of these possible outcomes will prompt us to reject the null hypothesis or, alternatively, to retain it. Note that in this context we never speak of proving a hypothesis, only of accepting or rejecting it as a tenable proposition. Our strategy will consist of partitioning the area of the sampling distribution into two parts: one a region of acceptance and the other a region of rejection. Then if our statistic, the sample mean, falls in the first of these regions, we shall accept the null hypothesis; if it is in the second region, we shall reject the null hypothesis.

Consider the problem of the trip distances. If our hypothesis is true, then the sampling distribution of the means will be as in Figure 4.4. The problem now is to decide upon the relative sizes of the regions of rejection and acceptance. If we make the region of rejection large, then we run the risk of rejecting a true hypothesis. This is known as a Type I error, and it is controlled by the selection of a probability value, α, the coefficient of risk. What is meant by this coefficient of risk? With reference to Figure 4.4, we

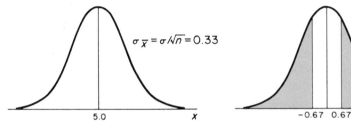

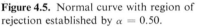

Figure 4.4. Sampling distribution of means.

Figure 4.5. Normal curve with region of rejection established by $\alpha = 0.50$.

know that even if 5.0 is the true population mean, sampling error alone will result in some sample means falling out in the tails of the distribution. The probabilities associated with these events, however, may be too low for us to accept the idea that their occurrence is consistent with the hypothesis of 5.0 being true. Our reaction in this situation would be to reject the null hypothesis and favor an alternative. Therefore, in one or both of the tails of the distribution we specify a region of rejection, the size of which is given by α, a probability value. For example, in Figure 4.5 the value of α is approximately 0.50, which is to say that the region of rejection, the shaded area, includes 50 percent of the area under the curve. In this case, there is obviously a greater risk of a Type I error than in the example of Figure 4.6, where α is only 0.05.

In actual research problems, the statement of alternative hypotheses and the location of the region of rejection in one or the other or both of the tails of the distribution is considered more carefully.

Aside from the Type I error, there is another possibility of error. This involves the acceptance of a false hypothesis, in other words, failing to reject H_0 when some other hypothesis is actually true. The probability of this Type II error depends upon α and the hypothesis H, which is actually true. Sup-

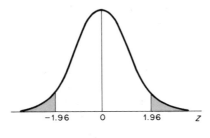

Figure 4.6. Normal curve with region of rejection established by $\alpha = 0.05$.

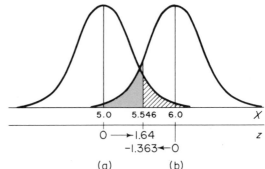

Figure 4.7. Diagrammatic representation of a type II error situation.

pose in our example, α was chosen as 0.05 and the corresponding region of rejection was located in the right-hand tail of the sampling distribution (Figure 4.7a). The decision to have only this "one-tail" test presumably stems from the substantive considerations of the research problem. In this case we are uninterested in the alternative hypothesis $\mu < 5.0$, but we are interested in the possibility that the true mean is actually greater than 5.0.

Assume, now, that the true mean is 6.0. If this is the case, the sampling distribution of the means will be as in Figure 4.7b. Then it can be seen that the probability of the sample mean falling in the region of acceptance, *given* that the true population mean is 6.0, is represented by the solidly shaded area.

This probability of a Type II error is computed as follows: First, for Figure 4.7a, we have to compute the *critical value* of $\bar{X}$, measured on the abscissa, which marks off the region of rejection. In terms of the Z scale, the critical value marking off an area of 0.05 probability to the right is given in tables of the standardized normal distribution function as approximately 1.64. Now by definition

$$Z = (X - \bar{X})/\sigma,$$

only in this case we are dealing with a distribution of mean values and, therefore,

$$Z = (\bar{X} - \mu)/(\sigma/\sqrt{n}).$$

In our example, we know the values of Z, μ, σ, and n, thus we can solve for $\bar{X}$:

$$1.64 = (\bar{X} - 5)/(\tfrac{2}{6});$$

therefore

$$\bar{X} = (1.64)(0.333) + 5$$
$$= 5.546.$$

Hence, the probability of a Type II error in the example is

$$P(\bar{X} \leqslant 5.546 \,|\, \mu = 6.0).$$

Now if we can express the value 5.546 as a Z score with respect to the mean of 6.0, then this probability also can be obtained from tables of the standardized normal distribution function.

$$Z = (\bar{X} - \mu)/(\sigma/\sqrt{n});$$

therefore

$$
\begin{aligned}
Z &= 5.546 - 6.0)/(\tfrac{2}{6}) \\
&= (-0.454)/0.333 \\
&= -1.363.
\end{aligned}
$$

From the tables we find that

$$P[Z \leqslant -1.363] \approx 0.09.$$

This probability of a Type II error often is represented as β and referred to as the "operating characteristic" of the test with respect to H.

An important related concept is the "power" of a test. This is the probability of rejecting H_0 when H is actually true, and it is equal to $(1 - \beta)$. A power curve, based upon several different alternatives for H, can be plotted for any test. Fraser (1958, pp. 248–250) discusses these concepts.

In many applied statistical problems, the trade-off between the relative probabilities of a Type I and Type II error is important; also, there is considerable attention given to the power of the tests employed, but in geographic research these issues have not been emphasized.

The steps involved in the testing of a statistical hypothesis may be summarized as follows:

(i)   The null hypothesis and the alternatives are stated clearly.

(ii)   A test statistic is chosen. The appropriate test statistics for most situations have been derived in the literature of mathematical statistics.

(iii)   A confidence level, or a coefficient of risk α, is selected.

(iv)   Assuming the null hypothesis is true, the appropriate sampling distribution is identified.

(v)   A region of rejection and the corresponding critical values of the test statistic are established, consistent with the value of α.

(vi)   The sample is chosen, and the statistic computed.

(vii)   Depending on whether the sample statistic falls in the region of acceptance or rejection, the null hypothesis is retained or rejected.

Nothing has been said about the question of sample size. It is possible to solve for the appropriate size of sample when certain confidence levels are specified for estimates or power levels are designated for tests. Krumbein and Graybill (1965, pp. 164–167) discuss some of the work on this question in geology; no such literature exists in geography, however. For the most part, only rough rules of thumb concerning the percentage of the population sampled, such as 10 or 25 percent, have been followed by geographers. From the point of view of much of the statistical theory, a sample of at least 30 observations usually is sufficient, providing the other formal assumptions of the theory are satisfied.

There is a large range of statistical hypotheses which can be tested. The appropriate tests are determined in part by the information which is known already concerning the population. In this section, the testing of some hypotheses as to the significance of regional differences is reviewed.

Comparison of Two Variances

In situations where there are two independent random samples, n_1 and n_2, drawn from normal populations, the significance of the difference between the two variances can be tested. The null hypothesis is that there is no significant difference and that the two populations are equally variable. The test statistic is the ratio of the two sample variances, s_1^2/s_2^2, which is distributed as F with $(n_1 - 1)$ and $(n_2 - 1)$ degrees of freedom. The larger of the two variances is always put in the numerator.

In studying the influence of lithology on slope steepness, Strahler (1954) sampled 100 points on each of two areas: one a region of shale the other a region of interbedded sandstones and shales. The sample information was as follows:

Shale	Sandstone-shale
$n_1 = 100$	$n_2 = 100$
$\bar{X}_1 = 31.82°$	$\bar{X}_2 = 33.12°$
$s_1 = 4.42°$	$s_2 = 5.78°$

$$F(99, 99) = \frac{5.78^2}{4.42^2} = 1.71.$$

For $\alpha = 0.05$, the region of rejection is approximately $F \geqslant 1.39$, and since the computed statistic falls in this region, the null hypothesis is rejected. The conclusion is that the difference in slope variability on the two rock types is statistically significant.

The test of equality of variance can be extended to the case of more than two samples. Bartlett's test is appropriate here and is discussed in Fraser (1958, pp. 260–261). The test statistic in this case is distributed as chi-square. King (1961) employed this test in studying the variability of distances between urban centers in different economic subregions.

Comparison of Two Regional Means

In testing the null hypothesis $H_0: \mu_1 = \mu_2$, with samples drawn from two normally distributed populations, the selection of a test statistic is dependent upon what is known about the two variances. If they are known and equal, then the test statistic is

$$Z = (\bar{x}_1 - \bar{x}_2)/(\sigma\sqrt{1/n_1 + 1/n_2}),$$

and the standardized normal distribution can be used. The same holds true if the variances are known but unequal. The only difference is that the two variances appear in the test statistic

$$Z = (\bar{x}_1 - \bar{x}_2)/(\sqrt{\sigma_1^2/n_1 + \sigma_2^2/n_2}).$$

If neither variance is known but they are assumed to be equal, then a "t" test applies. A pooled estimate of the assumed common variance is derived for this test as follows:

$$s_p^2 = [(n_1 - 1)s_1^2 + (n_2 - 1)s_2^2]/(n_1 + n_2 - 2).$$

The test statistic is then

$$t = (\bar{x}_1 - \bar{x}_2)/(s_p\sqrt{1/n_1 + 1/n_2}), \qquad \text{d.f.} = n_1 + n_2 - 2.$$

Again an example is taken from Strahler's review (1954) of quantitative geomorphic research. In a "badlands-type" area at Perth Amboy, New Jersey, a sample of 154 measurements of slope was made in 1948. The average slope was 49.1° and the standard deviation was 3.6°. Four years later, on the same area, at the same points, and with the same instruments, the measurements were repeated. For the 149 points which could be identified, the average slope was 48.8° and the standard deviation was 3.5°. The question was whether the difference in the means was significant. The population variances were unknown but assumed equal. The latter point could have been tested with regard to the sample variances, but they were obviously so close that the test was unnecessary. For the "t" distribution with degrees of freedom (d.f.) equal to $(n_1 + n_2 - 2) = 301$, the critical value of "t" at the level of $\alpha = 0.05$ is 1.97. The computed statistic was $t = 0.561$. In this case, the null hypothesis was retained, and Strahler concluded that "we have no reason to doubt that parallel slope retreat is occurring here."

A second example of this type of test is provided by Haggett (1965). For data concerning the spacing of towns in the two regions of eastern Texas and Iowa-Missouri, a test was made of the significance of the difference between the two mean values. The "t" statistic was computed as follows:

$$t = (20.5 - 19.2)/(43.4\sqrt{\tfrac{1}{67} + \tfrac{1}{203}})$$
$$= 0.219.$$

At the 95 percent confidence level ($\alpha = 0.05$), this obtained value of "t" resulted in the null hypothesis being retained. The difference between the mean values was judged to be not statistically significant.

Comparison of More Than Two Regional Means

The extension of the test of the significance of difference between two means to the case of more than two populations involves one of the more powerful forms of statistical analysis, the analysis of variance. The importance of this technique stems largely from its applicability in many problem situations and the fact that it can be structured in the form of several different models. Here we consider only a simple single-factor model. The reader is referred to Krumbein and Graybill (1965, pp. 197–218) for an excellent review of other forms of the analysis of variance as they are used in geology.

Assume that a variable is studied over a number of regions, to determine whether or not regional contrasts are significant. For example, to develop the problem mentioned earlier, it may be of interest to determine whether or not the average distance traveled by farm families for groceries differs significantly among regions in South Dakota, Iowa, Wisconsin, and Pennsylvania. The null hypothesis would be that

$$\mu_1 = \mu_2 = \mu_3 = \ldots = \mu_k,$$

for the set of k regions. An important assumption which has to be made in testing this hypothesis is that the k populations (regions) are equally variable with regard to the phenomenon under study. A preliminary test of this assumption can be made using Bartlett's test, which was mentioned earlier.

The analysis involves k random samples. As noted in Chapter 2, the total variation in this situation can be partitioned into two parts, a within-groups variation which relates to the deviations between observations and their respective sample means and a between-groups variation associated with the differences between the sample means and the grand mean for all observations. These different sources of variation usually are included in a summary of analysis of variance table (Table 4.1).

We recall that the populations are assumed to be equally variable. Therefore, if the ratio of the between-group and within-group variances is significantly large, it must be because the group means lie at different elevations.

The ratio of the between and within mean squares is distributed as F with $(k - 1)$ and $(N - k)$ degrees of freedom. The larger of the two mean squares appears in the numerator. A critical region is defined as before for a selected level of α, and if the computed value of F falls in this region, then the null hypothesis is rejected.

TABLE 4.1. SUMMARY OF ANALYSIS OF VARIANCE

Source of variation	d.f.	Sum of squares	Mean square
Within	$N - k$	$\displaystyle\sum_{j=1}^{k}\sum_{i=1}^{n_j}(X_{ij} - \bar{X}_j)^2$	$\dfrac{\text{SS within}}{N - k}$
Between	$k - 1$	$\displaystyle\sum_{j=1}^{k} n_j (\bar{X}_j - M)^2$	$\dfrac{\text{SS between}}{k - 1}$
Total	$N - 1$	$\displaystyle\sum_{j=1}^{k}\sum_{i=1}^{n_j}(X_{ij} - M)^2$	

The rejection of the null hypothesis and the acceptance of the alternative that not all the means are equal tells us nothing about the differences between particular pairs of the means. Scheffé (1959) discusses methods of handling this problem, however.

The analysis of variance discussed above involves a very simple model. It is this basic form of the technique, however, which has appeared in the geography literature, and one such example is discussed below.

In his study of Topeka urban land values, Knos (1962) employed the analysis of variance in testing a number of hypotheses. First he was interested in the possible effect of type of function upon land values. "Four categories of property; namely, single-family dwelling structures, multiple-family dwelling structures, industrial plants, and retail trade and service establishments were recognized. The assessed valuations for 70 properties were obtained and studied by means of an analysis of variance." The results, summarized in Table 4.2, prompted a rejection of the null hypothesis. A second hypothesis was that the mean land values would differ among

TABLE 4.2. RESULTS OF THE ANALYSIS OF VARIANCE BETWEEN
ASSESSED LAND VALUATION AND TYPE OF LAND USE,
TOPEKA, KANSAS, 1954

Source of variation	Sum of squares	Degrees of freedom	Mean square variance	F
Total variation	101,488,299,939	69		
Between-type variation	34,066,742,089	3	11,355,580,696	
				11.13*
Within-type variation	67,421,557,850	66	1,021,538,756	

*The least highly significant value of $F (P = 0.001)$ with 3 and 66 degrees of freedom is approximately 6.12.

Source: Knos (1962, p. 9).

three growth sectors of the city. Again the null hypothesis was rejected (Table 4.3).

Haggett (1964), it might be noted, employed a somewhat different analysis of variance model in his work on southeast Brazil. In studying the effects of different factors on the distribution of forest cover, he considered both the individual and joint effects of four main factors. A summary of his results is presented in Table 4.4.

In central place studies, Berry and Garrison (1958) and Mayfield (1967) used analysis of variance to confirm the existence of hierarchical classes defined with respect to sets of functions and population size.

TABLE 4.3. RESULTS OF ANALYSIS OF VARIANCE BETWEEN URBAN LAND VALUES AND SECTORS OF GROWTH, TOPEKA, KANSAS

Source of variation	Sum of squares	Degrees of freedom	Mean square variance	F^*
Total	124,851,668,348	74	1,687,184,707	
Between-sector	16,819,224,611	2	8,409,612,305	
				5.60
Within-sector	108,032,443,737	72	1,500,450,607	

*The least highly significant value of F ($P = 0.01$) with 2 and 72 degrees of freedom is 4.94.
Source: Knos (1962, p. 24).

TABLE 4.4. FORTALEZA BASIN: RESULTS OF MULTIPLE VARIANCE ANALYSIS OF FACTOR EFFECTS ON FOREST COVER

Nature of the effect	Source	Amount of the effect (angular units)	F ratio	Conventional significance level
Main factors:				
	Terrain (A)	+9.8	9.3	Probably significant*
	Soils (B)	+4.7	1.6	Not significant
	Farm size (C)	+4.0	1.2	Not significant
	Farm access (D)	+8.7	8.2	Probably significant*
Interactions:				
	BD (soils × farm size)		10.7	Probably significant*
	Other interactions between pairs		1.8	Not significant
	Higher-order interactions (ABC, ABD, . . . , ABCD)		. . .	. . .

*95 percent confidence level.
Source: Haggett (1964, p. 370).

Boyce (1965) has discussed the use of a "mixed model" analysis of variance for testing hypotheses concerning urban travel patterns. In his study, there were three fixed factors, trip direction, trip length, and size of destination zone, and one random factor, zone of origin. Three series of experimental designs involving these factors were applied to a sample of interzonal trip data, the criterion variable. The main hypothesis tested was that interzonal trip volume was independent of the direction of trips, and the results showed this to be a valid assumption for arterial trips but not for transit trips.

Goodness-of-fit Test

In the previous chapter it was noted that empirical estimates of the parameters in the probability laws are obtained by "fitting" the functions to sets of data. The question arises in these situations as to how good the fit is between the model and the data.

The fact that the statistic

$$\sum_{j=1}^{k} \frac{(f_j - F_j)^2}{F_j}$$

is distributed as chi-square with $(k - 1)$ degrees of freedom also has been noted. This is the test statistic in a commonly used goodness-of-fit test. The f_j are the observed frequencies, whereas the F_j are the expected or theoretical frequencies given by a particular probability function. The null hypothesis is that the differences between these two sets of frequencies are not statistically significant.

If we refer back to Table 3.2, we can note the chi-square values which Harvey obtained for his fits of the negative binomial. For each of the four chi-square values, the probability of it being equaled or exceeded over the long run was well above 0.05. In other words, at the 95 percent confidence level the null hypothesis was retained in each case. Harvey's (1966, p. 93) conclusion was as follows: "This interesting result suggests that the simulated pattern (*observed*) reflects very well the basic process of diffusion (*negative binomial*), but unfortunately the values in the cells are not independent and we cannot, therefore, infer that the simulated pattern is the same as that generated by a negative binomial model."

The chi-square test is not a particularly powerful one, and different results can be obtained by combining some of the classes. Usually it is desirable to have about 10 classes each with an observed frequency of at least 5, although there is no universal agreement on this requirement.

There are alternative tests which may be used in this context. The Kolmogorov-Smirnov test involves a comparison of the two cumulative distribution functions. Massey (1951) has tabled approximate significance levels for this test. Quandt (1964) reviewed the use of this test in a study of the discrimination among alternative hypotheses when fits of different probability laws are involved. The test did allow for a finer discrimination than would have been possible with a chi-square statistic, but the results were not convincing. Quandt proposed, therefore, a new statistic and two new associated measures of goodness of fit. The rationale for these is based on sophisticated notions of statistical analysis, however, which are not appropriately introduced here. The interested reader is referred to Quandt's article.

An important problem related to this question of fitting distributions to empirical data is that of testing whether a particular variable is normally distributed. Many theories of statistical inference demand this assumption of normality. There are a number of appropriate tests, but only two are mentioned here. The first is essentially a graphical one and is discussed by Hald (1952). It is based on the fact that the distribution function for the normal probability law graphs as a straight line on semilogarithmic paper. The line for a normal distribution having the same mean and variance as the observed data is plotted, and a selected confidence interval around this line also is graphed. Then the actual cumulative distribution points for the data are plotted and this provides the basis for accepting or rejecting, at the chosen level of confidence, the hypothesis that the observed data are normally distributed. The graphs involved are called *fractile diagrams*, and they have been used in geographic research by King (1961), Thomas (1961), and McConnell (1966b).

An alternative test considers the symmetry and shape of the observed distribution. It may differ from a normal distribution in its degree of *skewness* and/or its *kurtosis*, which relates to the proportion of observations concentrated near the mean. Snedecor (1956, pp. 199–202) discusses the appropriate measures of these two characteristics. Skewness is measured by the ratio of the third moment about the mean to the third power of the standard deviation. The measure of kurtosis involves the ratio of the fourth moment about the mean to the fourth power of the standard deviation. In both cases, the deviation of the shape factor from zero is a measure of the departure from normality. The statistical significance of these departures is tested using a "t" statistic.

Berry (1963, p. 230) ran these tests on the 10 variables used in his classification of commercial centers in Chicago. Logarithmic transformations were applied subsequently to ensure that all the variables were normally distributed.

4.5 NEW TRENDS IN STATISTICAL INFERENCE

The discussion of statistical inference in this chapter perhaps may appear longer than should be justified by the limited range of applications of the methods in geographic research to date. Despite its length, however, the treatment has been oversimplified and has ignored some important topics in statistical inference. Three of these topics are introduced briefly in this section.

The first concerns the use of *order statistics*. These relate to values which are arranged in order of size or magnitude. Consider an independent random sample $(x_1, x_2, x_3, \ldots, x_n)$ drawn from a continuous distribution. The latter assumption usually is emphasized in most discussions of the related theory. The sample observations $(x_1, x_2, x_3, \ldots, x_n)$ can be considered as independent random variables. Assume that they are now ordered in size and a new set of variables $(x'_1, x'_2, x'_3, \ldots, x'_n)$ is introduced to represent these ordered values, such that

$$x'_1 < x'_2 < x'_3 < \ldots < x'_n.$$

Then it is of interest to derive the probability distributions for certain of these variables, either singly or in combination, and the sampling distributions for statistics based on them. For example, it is possible to obtain the probability distributions for the extreme values or for the median or the range. Johnson and Leone (1964, Chapter 6) outline a simple introduction to this topic. Wilks (1948) provides a more extended review.

In his work on point processes, Dacey (1963, 1965a) treats the distances between neighboring points as order statistics. For any point *i*, the statistics

$$r_{i1} \leqslant r_{i2} \leqslant r_{i3} \leqslant \ldots \leqslant r_{ik}$$

are the order statistics corresponding to the ordered distances from the point to its first, second, third, and, finally, *k*th nearest neighbor. The probability distribution for these statistics is derived.

The question of inferential problems in map pattern analysis will be discussed more fully in the next chapter, and it is not pursued at this point.

It is important to note that order statistics are not the same as rank order statistics. With the latter there is a rank value $(1, 2, 3, \ldots, n)$ substituted for the original sample value. This is not the case with order statistics, where the sample values are retained but in an ordered arrangement.

Rank order statistics are important in many *nonparametric* techniques of statistical inference. This is the second topic which is to be introduced in this section, namely nonparametric tests. These are distinguished usually by the fact that they do not specify any parameters and, related to this, by the fact that they do not assume any particular form of a distribution from which the sample is drawn. In the "classical" approach to statistical inference discussed earlier, much of the theory is derived on the assumption that normal probability distributions are involved. Although certain tests are robust and not very sensitive to this assumption being violated, it often is preferable to use a *distribution-free* test in such situations.

The terms *nonparametric* and *distribution-free* generally are used interchangeably, for instance, Fraser (1958, p. 378). Johnson and Leone (1964, p. 259), however, point out that "distribution-free" is a more demanding requirement than "nonparametric," and while both features may apply, it is not always so. As an example, they note that tests based on signs are largely distribution-free but not necessarily nonparametric.

It is impossible to summarize here all the nonparametric methods. Hoel (1960, Chapter 11), Fraser (1958, Chapter 16), and Johnson and Leone (1964, Chapter 9) give introductions to the subject. Siegel (1956) and Fraser (1957) discuss the different methods in considerably more detail. The interested reader should consult these sources.

In geography the techniques have not been used widely. Reference has been made already to the Kolmogorov-Smirnov test and its possible applications. Haggett (1965, pp. 292–293) illustrates the use of the Mann-Whitney U test in considering the relationship between a variable which is represented by groups and another for which there is rank-order data. Thompson, et al. (1962) used rank correlation methods in measuring the interrelationships among a number of variables concerned with economic health. Sternstein (1962) has argued for even greater use of these rank correlation methods in geographic research.

The third and final topic to be introduced is a more controversial one around which considerable debate still is going on in the field of statistics. It is subjective probability or Bayesian analysis. The two names are used at different times, apparently without unanimous agreement as to which is the more appropriate one.

The approach makes use of an old theorem of conditional probability, Bayes's theorem. Following Goldberg (1960, pp. 91–98), this may be stated as follows: Given a set of events $(E_1, E_2, \ldots, E_n)$, a partition of the sample space (each with a nonzero probability), and E as any event (also with nonzero probability), then for the integer k, where $1 \leqslant k \leqslant n$,

$$P(E_k \mid E) = P(E_k) \cdot P(E \mid E_k) / \sum_{j=1}^{n} P(E_j) \cdot P(E \mid E_j).$$

Goldberg notes that the left-hand side often is called the *a posteriori probability* of the hypothesis E_k, given the observed event E, whereas $P(E_k)$ is the *a priori probability* of the hypothesis E_k.

As Plackett (1966) has observed, this theorem is true regardless of one's interpretation of probability. For example, following the definitions of probability given earlier in this book, one could use Bayes's theorem as long as a relative frequency definition of the a priori probability did exist. But such knowledge of the prior distribution is not always obtainable. As a consequence, the name *Bayesian analysis* is best applied in statistical inference to those approaches which in Plackett's words (1966, p. 250) view probability as "a numerical measure of degree of belief" and which use Bayes's theorem "in order to convert prior degrees of belief, taken with the evidence supplied by the data, into posterior degrees of belief."

Hodges and Lehmann (1965, pp. 112–116) provide some simple examples of this approach in statistical analysis. Plackett (1966) and Hirshleifer (1961) give more extended treatments of the Bayesian approach in the context of testing hypotheses and estimation, and Plackett particularly references many of the important works on the subject.

The only published reference to Bayesian analysis in the geography literature is by Curry (1966a). In reviewing the general question of decision making in the resources field, Curry identifies some interrelated problems which might be handled in a Bayesian framework. First, there is the seasonal programming problem faced by a farmer. He has a prior "probabilistic forecast of rainfall, representing his degree of belief that certain values of rain will occur, the odds he will accept by committing himself to a program that assumes certain rain values" (Curry 1966a, p. 135). The results of the rain-seeding experiments are the observed events which, combined with the prior probabilities, yield posterior odds. The farmer weights these odds by his calculated utilities and evaluates both the seeding experiments and his farm program accordingly.

The problems of seasonal forecasting and weather modification are structured in a similar framework by Curry.

It is possible that questions of Bayesian analysis will be emphasized more in the geography literature in the future. On one hand, it may be relevant, as Curry suggests, in the context of trying to understand the decisions made by man with respect to spatial problems. On the other, it may provide a more meaningful framework in which to test hypotheses concerning spatial parameters and processes.

SUGGESTED READINGS

Berry, B. J. L. (1962), "Sampling, Coding, and Storing Flood Plain Data." *Agricultural Handbook*, United States Department of Agriculture, No. 237.

Boyce, D. E. (1965), "The Effect of Direction and Length of Person Trips on Urban Travel Patterns." *Journal of Regional Science*, Vol. 6, pp. 65–80.

Curry, L. (1966a), "Seasonal Programming and Bayesian Assessment of Atmospheric Resources." In W. R. D. Sewell (ed.), *Human Dimensions of Weather Modification. Research Paper 105*, Department of Geography, University of Chicago, pp. 127–138.

Fraser, D. A. S. (1958), *Statistics: An Introduction*. New York: John Wiley & Sons, Inc., Chapters 6–11.

Haggett, P. (1965), *Locational Analysis in Human Geography*. London: E. Arnold, Ltd., pp. 277–303.

Hoel, P. G. (1960), *Elementary Statistics*, Second edition. New York: John Wiley & Sons, Inc., Chapters 5–7.

Quandt, R. E. (1964), "Statistical Discrimination Among Alternative Hypotheses and Some Economic Regularities." *Journal of Regional Science*, Vol. 5, pp. 1–23.

Strahler, A. N. (1954), "Statistical Analysis in Geomorphic Research." *Journal of Geology*, Vol. 62, pp. 1–25.

INFERENTIAL PROBLEMS
IN THE ANALYSIS
OF LOCATION PATTERNS

This chapter considers applications of statistical techniques in the description and analysis of spatial distributions as represented in mapped location patterns. The question of testing hypotheses about the nature of spatial patterns forms the major part of this discussion.

5.1 DEFINITION OF A LOCATION PATTERN

To begin with, there is a need for elaboration on what is meant by the phrase "a location pattern." Geographers usually rely upon two operational definitions of this concept. On one hand, they treat locations as *points* on a map and analyze the distances separating points, the densities of points in areas and subareas, the distribution and arrangement of points among subareas, and the degree of correspondence between point patterns. More recently, this approach has been

5

extended to include the discussion of theoretical point patterns, to which some reference already has been made in our earlier discussion of probability laws. It was noted that the appropriate probability functions have been derived for both homogeneous and inhomogeneous random point patterns and for patterns more uniform than random. In the present chapter, these theoretical patterns are considered in somewhat more detail, especially in regard to the distances between the points contained in the patterns. The problem of comparing distance measures for observed point patterns with these theoretical values also is emphasized. The discussion of point pattern analysis in general will point up the fact that this approach to location analysis has been most fruitful in terms of analytical results, but the attempts at providing empirical interpretations of the results have not been equally successful.

An alternative approach in the analysis of a location pattern is by way of using *areal units* such as grid squares or county units. In much of the work already accomplished along these lines, the statistical analysis appears to have supplemented the established choropleth method of map analysis, and the results have not been impressive. Apart from the fact that the observations in the associated data matrices are areal units, much of this analysis has been essentially nonspatial with little to distinguish it from the analysis completed in other disciplines. This is not to overlook the fact that there are statistical techniques common to all fields of research, but only now is it becoming clear that the description and analysis of location patterns in terms of a set of areal units poses some peculiar and challenging statistical problems still to be resolved. The derivation of an appropriate measure of contiguity is one such problem which is introduced in this chapter.

In the geographer's discussion of location patterns, a distinction often is made between spatially *discrete* and spatially *continuous* phenomena. The former are defined only at certain points on the earth's surface, as metropolitan areas or manufacturing plants are, whereas continuous phenomena, for example, elevation, temperature, or pressure, are defined at every point. In most cases this is a quite logical distinction, but depending on the *scale* of the investigation, it may prove quite misleading. On a world scale, "metropolitanization" might be said to be a discrete phenomenon, but in the megalopolis area of the northeastern United States it might be regarded quite properly as a continuous phenomenon. Indeed, the distinction between discrete and continuous phenomena is best defined in the more formal mathematical sense, discussed in Chapter 3 on probability.

In the following two sections of this chapter, the description and analysis of point patterns in terms of distance measures are discussed. Sections 5.4 and 5.5 reflect the distinction which Dacey (1966b) has emphasized about the analysis of point patterns, namely, the difference between *distribution* and *arrangement*. The tests which are applicable in studying these two different aspects are outlined in these sections. The final section deals with some questions relating to location patterns other than point patterns.

5.2 DISTANCE MEASURES FOR THE DESCRIPTION OF OBSERVED POINT PATTERNS

In this section we consider distance measures which are useful in describing observed point patterns.

Nearest-neighbor Measures

The locations of points relative to one another, and the distances between nearest neighbors, are important expressions of arrangement in a point pattern.

Plant ecologists, such as Clark and Evans (1954), Morisita (1957), and Pielou (1959, 1961), first developed these methods for the analysis of plant communities. These contributions are reviewed excellently in Greig-Smith (1964). The early geographic work along these lines, for example, Dacey (1960b) and King (1962a), relied essentially upon techniques borrowed from this plant ecology literature. More recently, however, some formal extensions of this type of analysis have been outlined in the geographic literature, and it is these contributions which are emphasized here.

Consider the point pattern in Figure 3.5 which was discussed earlier. Let i be any point and r_{ij} the distance between i and the jth nearest point. If there are j measurements made from i, and these measurements are ordered such that

$$r_{i1} \leqslant r_{i2} \leqslant r_{i3} \leqslant \ldots \leqslant r_{ij},$$

then r_{ij} is referred to by Dacey (1963) as the jth-order distance. Dacey notes that "for most practical applications, measurements from i should be recorded only to those j neighbors which are closer to i than i is to the boundary." Also, to eliminate the effects of map scale, it may be desirable to standardize the r_{ij} by multiplying them by $d^{1/2}$, where d is the density of points in the study area.

Assuming n points in the area, then the j mean-order distances can be derived as follows:

$$\bar{r}_j = \frac{1}{n} \sum_{i=1}^{n} r_{ij}.$$

Dacey (1963) has generalized these concepts of ordered distances to nearest neighbors to handle the description of patterns in multidimensional space. The two-dimensional case, as represented by a map, however, is of sufficient interest in this discussion.

The nonstandardized mean values for the pattern in Figure 3.5, with $j = 1, 2, 3, 4$, are presented in Table 5.1. The variances of the four order statistics also have been computed.

TABLE 5.1.   ORDER DISTANCE STATISTICS* FOR PATTERN IN
FIGURE 3.5.

Order j	n	$\sum_{i=1}^{n} r_{ij}$	$\bar{r}_j$	σ_j^2
1	34	13.13	0.39	0.03013
2	28	15.49	0.55	0.04095
3	25	17.52	0.70	0.03292
4	21	16.85	0.80	0.03129

*Measurements in miles.

These observed distance statistics, particularly the mean distances, then can be used in pattern interpretation. As we shall discuss in Section 5.3, they usually are compared in an inferential framework with the parameters for particular theoretical patterns.

Regional Method of Nearest-neighbor Analysis

An alternative approach to the description of point patterns is the "regional method." This is discussed by Dacey and Tung (1962). The map space around each point is divided into k equal-sized sectors or regions. The distance from the center point to the nearest point in each sector is measured, and "these observed distances are used to classify the regions in some systematic fashion and a separate mean is obtained for each kind of region" (Dacey and Tung, 1962, p. 85). It is common to use six regions, as illustrated in Figure 5.1. For point i at the center, r_{i1} is the shortest of the six measurements, r_{i2} the second shortest, and so on. Thus,

$$r_{i1} \leqslant r_{i2} \leqslant \ldots \leqslant r_{i6}.$$

Then

$$\bar{r}_{ik} = \left(\sum_{i \in I} r_{ik} \right) \Big/ n,$$

where I is the set of n points for which measurements are made. It may be the case that all the points in the pattern are in I or, alternatively, that there may be only a sample of n of these points. In either case, there are some practical problems to be resolved. Dacey and Tung observe that there may be some of the points, especially near the boundary, which have regions

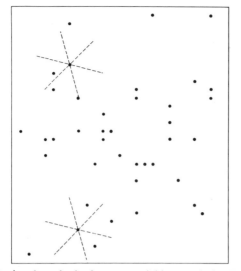

Figure 5.1. Regional method of nearest-neighbor analysis. For illustration purposes the sectors are drawn in for only two of the points.

containing no neighbors. These points may be ignored completely, although Dacey and Tung prefer to include them and to work with a variable number of observations in computing each regional mean. The latter procedure has been used in computing the mean values for the pattern in Figure 5.1. The corresponding values are given in Table 5.2.

TABLE 5.2.   REGIONAL MEAN DISTANCES FOR FIGURE 5.1*.

k region	n points	$\sum_{i=1}^{n} r_{ik}$	$\bar{r}_{ik}$
1	34	13.29	0.39
2	31	18.25	0.59
3	24	17.83	0.74
4	20	18.05	0.90
5	17	19.72	1.16
6	9	12.11	1.36

*Measurements in miles.

Again these regional statistics for observed point patterns can be compared with the values for theoretical patterns; this is noted shortly.

In comparing the order method and regional approach to point pattern analysis, Dacey and Tung (1962, p. 86) note that while "the relative power efficiency of the two nearest-neighbor approaches has not been accurately evaluated, the regional method appears to have greater efficiency for discerning random patterns and for detecting the presence of randomness in a systematic pattern biased in the direction of uniform spacing. On the other hand the regional method has less efficiency than the order method when the locations are biased toward grouping or clustering."

Bachi's Standard Distance Statistic

Bachi (1963) has suggested, as a measure of the dispersion of a set of points, a "standard distance," d_{ij}, measured with respect to two orthogonal axes. The expression is

$$d_{ij} = [(x_i - x_j)^2 + (y_i - y_j)^2]^{1/2}.$$

This is the formula for the distance between two points in Euclidian space (Figure 5.2). For a set of n points, n^2 of these distances can be computed, assuming the distance from each point to itself is included, and every distance d_{ij} is measured twice, once for the point i and again for the point j. As a summary statistic for this set of n^2 distances, Bachi suggests "the mean quadratic distance," defined as

$$D = \left[\left(\sum_{i=1}^{n} \sum_{j=1}^{n} d_{ij}^2 \right) \Big/ n^2 \right]^{1/2}.$$

Then, by introducing the concept of the "mean center of the distribution" having the coordinates

$$\bar{x} = \left(\sum_{i=1}^{n} x_i \right) \Big/ n, \quad \text{and} \quad \bar{y} = \left(\sum_{i=1}^{n} y_i \right) \Big/ n,$$

Bachi shows that

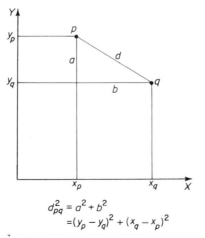

$$d_{pq}^2 = a^2 + b^2$$
$$= (y_p - y_q)^2 + (x_q - x_p)^2$$

Figure 5.2. Distance between two points in Euclidean space.

$$D = \left(\frac{2 \sum_{i=1}^{n} d_{ic}^2}{n} \right)^{1/2}.$$

The term d_{ic}, the distance from point i to the mean center c, equals

$$[(x_i - \bar{x})^2 + (y_i - \bar{y})^2]^{1/2}.$$

Finally, the expression

$$d = \left[\frac{\sum_{i=1}^{n} d_{ic}^2}{n} \right]^{1/2}$$

is defined as the "standard distance."

If the map area is divided into k regions, then the partitioning of total variance discussed earlier in Chapter 2 can be applied in the case of the standard distance. Bachi notes that the square of this distance (d^2) for the whole area is equal to the sum of a "within-regions" squared distance (d_w^2) and a "between-regions" squared distance (d_b^2).

That is,

$$d^2 = d_w^2 + d_b^2$$

$$= \frac{\sum_{j=1}^{k} \sum_{i=1}^{n_j} [(x_{ij} - \bar{x}_j)^2 + (y_{ij} - \bar{y}_j)^2]}{n} + \frac{\sum_{j=1}^{k} n_j [(\bar{x}_j - \bar{x})^2 + (\bar{y}_j - \bar{y})^2]}{n},$$

where (x_{ij}, y_{ij}) are the coordinates of the ith point in the jth region, $(\bar{x}_j, \bar{y}_j)$ is the regional mean center, and $(\bar{x}, \bar{y})$ is the mean center for the whole distribution. The ratio (d_b^2/d^2) provides a crude measure of the significance of the regional divisions in the pattern, for if the ratio is close to 1, then something approaching maximum differentiation exists; whereas a ratio value close to zero suggests that the regions have the same mean center as the total pattern does.

Mean and Median Centers of Point Patterns

The concept of the mean center of a two-dimensional distribution pattern to which Bachi refers has been discussed in the literature of geography and other disciplines for some time (Hart, 1954). Seymour (1965) provides a computer algorithm for locating this center. The problem is handled as follows:

"The bivariate mean of an arbitrary number of points is located by obtaining the arithmetic mean of the x-coordinates and the arithmetic mean of the y-coordinates of all the data points. The resulting means are the coordinates of the bivariate mean" (Seymour 1965, p. 2). The corresponding formulae are identical with those in Bachi's analysis:

$$\bar{x} = \frac{1}{n} \sum_{i=1}^{n} x_i, \qquad \bar{y} = \frac{1}{n} \sum_{i=1}^{n} y_i.$$

Seymour notes that appropriate weightings also can be associated with each point.

The problem of locating the median center is a more difficult one since, as Seymour (1965, p. 4) points out, "one can locate the bivariate median to whatever precision the computer will allow." The use of grid overlays and an iterative procedure, however, allows for an approximate solution to be obtained. The following steps are involved:

(i)   A coordinate grid with equal numbers of horizontal and vertical axes is overlaid on the map. The limits of this grid are given by the locations of the four most extreme points in the pattern (Figure 5.3).

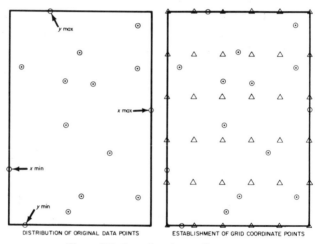

DISTRIBUTION OF ORIGINAL DATA POINTS   ESTABLISHMENT OF GRID COORDINATE POINTS

Figure 5.3. Locating the median center.

(ii)   For each new grid coordinate point (x_0, y_0), the square root of the sum of the squared distances to the n points (x_i, y_i) of the original pattern is calculated. The corresponding expression is

$$\left\{ \sum_{i=1}^{n} [(x_0 - x_i)^2 + (y_0 - y_i)^2] \right\}^{1/2}.$$

The grid point (x_j, y_j) having the lowest value for this expression is identified.

(iii)   This point (x_j, y_j) is taken as the center of a new but finer grid overlay which will encompass now only a subset of the original observation points. The number of gridpoints, however, remains constant. The fineness of this grid and the associated rate of areal reduction are determined by the researcher. The distance expression above is computed for the new grid points and a new point of minimum aggregate travel (x_k, y_k) is identified.

The third step is repeated as many times as necessary, given the particular degree of refinement desired. Seymour cautions that too many iterations only produce a "false accuracy" in that the location of the data points themselves will be subject to some error. This establishes presumably a lower limit on the accuracy of the median computation.

In two-dimensional space the median can be considered as the point of minimum aggregate travel and, as such, it has been the topic of considerable discussion in geography and related fields. Porter (1963, 1964) and Court (1964) recently have emphasized the indeterminancy of the solution to this problem.

Warntz and Neft (1960) also have proposed a number of distance concepts for spatial analysis. For the most part, their measures are based on the notion of an infinitesimal element of the map area within which there is a particular density of the pattern elements. Then, by use of simple calculus, expressions are derived for various "average" centers and measures of dispersion. For example, the mean center is that point at which $\int r^2 D\, dA$

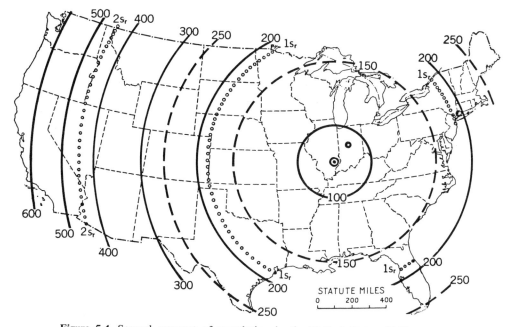

Figure 5.4. Second moment of population in the United States, 1950. Shown here are contours in units of trillions, persons times miles-squared. This is the physical definition of moment. To obtain the value for the second statistical moment at any point, as defined in the text, the value interpolated from the map must be divided by the total population. Exhibited also are the mean center, median center, and the modal center in Illinois, Indiana, and New York City, respectively. Shown, too, are segments of circles with the mean center as the origin and with radii of one and two standard deviations. *Source:* Warntz and Neft (1960).

is minimized. In this integral expression dA is the small element of area, D is the density for that area, and r is the distance from each area to the point concerned. Similarly, the point at which $\int rD\,dA$ is minimized defines the median center.

In discussing measures of dispersion, Warntz and Neft suggest that the moments about any point in a pattern can be derived by the formula

$$M_n = \frac{\sum (pr^n)}{P},$$

where n is the particular moment concerned, p is the population within a segment of area, r is the distance from the midpoint of that segment to the particular point, and P is the total population. A contour map of the second moment about the mean center for the 1950 population of the United States is shown in Figure 5.4.

In a recent monograph, Neft (1967) has developed this approach considerably. The interested student should consult this more detailed work.

Concept of Potential

A related distance concept which Warntz and others have emphasized is that of the "potential" at any point in a pattern. This potential model and the related "gravity model" appear in many forms in the social science literature, but we are concerned here only with the distance formulation. The potential at any point is defined as

$$\int \frac{1}{r}\,D\,dA,$$

where the terms are the same as before. For practical purposes, however, the expression

$$\left(\sum_{i=1}^{A} p_i\right)\Big/r$$

is preferred, where A is the total number of small subareas involved, p_i is the population in each subarea, and r is the distance from the particular point to a representative point in each subarea. A "potential-of-population" map is shown in Figure 5.5.

Court (1966) has drawn attention to the fact that usually some very simplifying assumptions are made in computing the contribution of an area's own population to the total potential existing at the central point of the area. For example, in the metropolitan areas of New York and Chicago and in the large, sparsely settled western counties, the population was assumed by Warntz to be distributed uniformly. Court, however, suggests that other probability models for the distribution of population might be more realistic

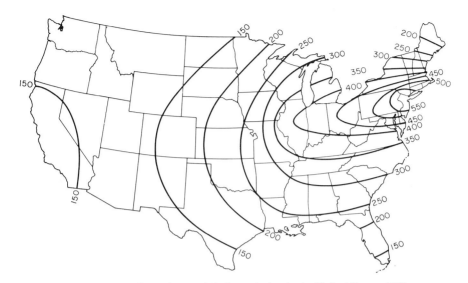

Figure 5.5. Density and potential of population in the United States, 1950. The contoured values are in units of thousands, persons per mile. The value at any point is expressed by the formula $\int (1/r)\, D\, dA$. Values greater than 550 are not shown. Within the 550 contour there is a steep gradient to approximately 1000 as the base potential from which the New York City peak rises. *Source:* Warntz and Neft (1960).

than this uniform distribution. His own computations prompted the conclusion that

> The self-potential imposed by a population uniformly distributed over a circle is only $\frac{2}{3}$ of that imposed by a population having either a conical or a Gaussian distribution, and less than half that imposed by one having a negative exponential distribution; the latter two are evaluated on the assumption that 95 percent of the population is inside the circle of radius c. These differences require careful examination of the distribution to be used in any calculation of self-potential (Court 1966, p. 29).

5.3 DISTANCE MEASURES FOR HYPOTHETICAL PATTERNS AND RELATED INFERENTIAL QUESTIONS

It is often of interest in geographic research to consider whether or not an observed spatial distribution conforms to a particular theoretical model. Some questions along these lines which emphasize distance measures are reviewed below.

Randomness of Points Along a Line

Consider a locational arrangement of points along a line, for example, the intersections along a major highway (Figure 5.6). It might be hypothesized that this pattern is a random one.

There are a number of statistics which can be used to test for randomness in this situation. The reader is referred to Barton and David (1956) and Pearson (1963) for a discussion of some of these tests. The sampling distributions for many of these statistics are not widely available in tabled form however; hence this discussion refers to only one such statistic which is normally distributed.

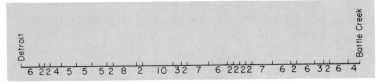

Figure 5.6. Distribution of interchanges along Interstate 94 in Michigan. Distances shown are in miles.

For a set of $(n - 1)$ points distributed along a line, let x_i be the distance of the ith point from one end of the line or the origin. It is convenient to assume that the line has *unit* length. These distances are ordered such that

$$x_1 \leqslant x_2 \leqslant x_3 \ldots \leqslant x_{n-1}.$$

Now define d_i as the interval separating two adjacent points. Then there will be n such intervals with

$$d_1 = x_1,$$
$$d_i = x_i - x_{i-1}, \qquad 1 < i \leqslant n - 1$$

and

$$d_n = 1 - x_{n-1}.$$

The statistic

$$Y = \sum_{i=1}^{n} i d_i$$

can be derived, and for $n \geqslant 25$ it can be shown that $Y' = 1 - Y/n$ is normally distributed with mean $(n - 1)/2n$ and variance $(n - 1)/12n^2$. The application of this test to the situation represented in Figure 5.6 is shown in Table 5.3. The corresponding normal deviate is extremely large, and the null hypothesis of randomness is rejected.

TABLE 5.3.   TEST OF RANDOMNESS FOR LINEAR PATTERN OF
POINTS, FIGURE 5.6.

	Figures in miles		
i	d_i	i	d_i
1	6	15	6
2	2	16	2
3	2	17	2
4	4	18	2
5	5	19	2
6	5	20	7
7	5	21	6
8	2	22	2
9	8	23	6
10	2	24	3
11	10	25	2
12	3	26	6
13	2	27	4
14	7		

Number of points $= 26$
Number of intervals $(d_i) = 27 = n$

$$Y = \sum_{i=1}^{n} id_i = 1564$$

$$Y' = 1 - Y/n = 1 - 57.92$$
$$= -56.92$$

Mean $Y' = \dfrac{n-1}{2n} = 0.48$

$$s'_y = [(n-1)/12n^2]^{1/2} = 0.0545.$$

Randomness of Points in Two-dimensional Patterns

The approaches to the problem of testing for randomness which are discussed here focus upon the nearest-neighbor distance. There are two different tests presented.

The first test is based on the assumption that the first nearest-neighbor distances in a random point pattern are normally distributed.

Assuming the n points in an area are distributed randomly in accordance with a Poisson probability function with density λ, and the distribution of distances between points and their first nearest neighbors is normal, then the expected value for these distances $(\bar{r}_E)$ will be

$$\bar{r}_E = 1/(2\lambda^{1/2}).$$

The standard error of this mean distance is given as $\sigma_{\bar{r}_E} = 0.26136/(n\lambda)^{1/2}$. The statistic $c = (\bar{r}_0 - \bar{r}_E)/\sigma_{\bar{r}_E}$, where $\bar{r}_0$ is the observed mean nearest-neighbor distance, is distributed as a normal variate and is used in the test of randomness. It should be stressed that $\bar{r}_0$ and λ have to be in the same metric.

We note at this point that the ratio $R = \bar{r}_0/\bar{r}_E$, defined as the *nearest-neighbor measure*, provides an index of the departure from randomness. The ratio is less than, equal to, or greater than *one*, depending upon whether the pattern tends to be aggregated, random, or uniform, respectively.

Getis (1964) employed this test in his study of the changing patterns of retail establishments in Lansing, Michigan, which was referred to earlier in this book. The statistics which he computed for the different years and the corresponding patterns of N stores are summarized in Table 5.4. The

TABLE 5.4.   SUMMARY OF NEAREST-NEIGHBOR MEASURES*

Year	N	p	$\bar{r}A$	$\bar{r}E$	R	$\bar{r}_E$	z
1900	20	0.000000594	696.29	648.50	1.074	75.82	−0.63
1910	33	0.000000899	354.87	527.31	0.673†	47.98	+3.59
1920	94	0.000000918	343.24	521.87	0.658†	28.14	+6.35
1930	124	0.000001097	368.33	477.37	0.772†	22.11	+4.93
1940	133	0.000000956	405.05	511.35	0.792†	23.18	+4.59
1950	117	0.000000797	470.86	560.10	0.841†	27.06	+3.30
1960	68	0.000000375	814.46	816.46	0.998	51.75	+0.04

*Values in feet.
†Significant at the 0.01 level.
Source: Getis (1964, p. 395).

same test of randomness was employed by King (1962a) in his study of the spacing of towns in selected areas of the United States, and it is implicit in Medvedkov's (1964) discussion of the same problem.

Nearest-neighbor analysis using normal probability theory has been generalized by Dacey (1963) in two respects. One extension relates to distances to other than the first nearest neighbor. For a point i, distances can be measured to different order nearest neighbors and the test for randomness applied to any set of jth-order distances.

The second extension is a generalization of the test for k-dimensional space. In this section we have been concerned only with points in a plane, that is, in two-dimensional space, but the analysis is easily extended to three-, four-, or k-dimensional space.

An alternative test of randomness for point patterns makes use of a chi-square statistic. Dacey (1963, 1964a) derived this test by showing that for a homogeneous random point pattern obeying a Poisson law, the distance from a randomly selected point to its jth nearest neighbor is described by a gamma probability law with parameters r and λ both greater than zero. That is,

$$f(r_j) = [2(\pi\lambda)^j r^{2j-1} e^{-\pi\lambda r^2}]/(j-1)! \qquad j = 1, 2, 3, \ldots,$$
$$= 0 \qquad\qquad\qquad\qquad\qquad \text{otherwise.}$$

The parameter λ is the density of points per unit area.

Given this result, it then can be shown that the statistic

$$2\pi d^{1/2} \sum_{i=1}^{n} r_{ij},$$

where n is the number of points in the pattern, d is the observed density, and r_{ij} is the jth-order nearest-neighbor distance for locus i, is distributed as chi-square with $2jn$ degrees of freedom.

Some of Dacey's results for the location pattern of towns in Iowa are given in Table 5.5. The fact that n decreases as the order j increases is related

TABLE 5.5.   STANDARDIZED, OBSERVED, AND EXPECTED ORDER
DISTANCES AND THE ORDER CHI-SQUARE STATISTIC FOR
DISTANCES FROM URBAN PLACES IN IOWA

j	n	$E(r_j)$	$\bar{r}_j$	$2\pi d^{1/2} \sum r_j^2$	$N = 2jn$	Standard units
1	65	0.5000	0.65	200.8889	130	4.00
2	64	0.7500	0.84	300.6266	256	1.92
3	58	0.9375	0.99	358.6284	348	0.42
4	52	1.0937	1.12	423.8500	416	0.29
5	47	1.2305	1.24	464.4787	470	−0.16
6	41	1.3535	1.36	499.9957	492	0.27
7	40	1.4663	1.49	567.4058	560	0.24
8	39	1.5710	1.60	638.4762	624	0.42
9	38	1.6692	1.68	686.2364	684	0.11
10	37	1.7620	1.78	741.3591	750	0.05

Source: Dacey (1963, p. 507).

to the "border effect" mentioned earlier. A measurement to a jth-order nearest-neighbor is discarded if the map border is closer to the locus point i than is the neighbor in question.

A normal approximation can be used in this situation. In Table 5.5 there are listed the corresponding standardized normal variates, $[(2\chi^2)^{1/2} - (2n-1)^{1/2}]$. For a coefficient of risk $\alpha = 0.05$, the null hypothesis that the location pattern of towns in Iowa is random was rejected for the first-order nearest neighbor but accepted for the higher orders. Dacey (1963, p. 511) concluded that "this test suggests the Iowa map pattern of urban places contains a large amount of randomness but the pattern is not properly or adequately described as random."

In concluding this section, it is as well to reiterate the assumptions upon which the tests above are based and at the same time to note their limitations.

First, the tests are sensitive to nonrandomness, but they do not allow for the evaluation of any alternative hypotheses such as uniformity or concentration of the points in clumps. In this regard, we might note that the probability functions and the corresponding moments for nearest-neighbor distances in point patterns other than a homogeneous random one already have been derived for a number of cases. Dacey (1965a) gives these results for an inhomogeneous random point pattern. A subsequent set of three studies (Dacey 1965c, d, e) reports on results obtained for

(i)   The distance from a point randomly located in a polygon to the *i*th nearest corner of a square, diamond, and rectangle.

(ii)   The distance between two points randomly and independently placed in a line segment, disk, sphere, or rectangle.

(iii)   The distance from a selected point to the nearest point in a collection of $(n + 1)$ points randomly and independently placed in a square.

(iv)   The relations between the moments of distance for random arrangements of points in a unit disk compared to a Euclidean plane and a unit area square compared to a Euclidean plane.

In the following subsection we shall discuss a study which develops some of this more advanced work, particularly that represented under the first of the four topics above.

Another shortcoming which should be mentioned is that many of the results in this whole section, including the tests for randomness, are derived on the assumption that the area in which the points are located is unbounded. Dacey (1965d) has discussed the possible consequences of the boundary bias in the analysis of spacing, and he notes (1965d, p. 9) that "nearest-neighbor measurements from the bounded random pattern (*a unit square*) are substantially greater than measurements from the unbounded random pattern where the unit square contains a small number of points." A topological solution to the problem of boundary effect is suggested by Dacey. It involves mapping the points onto a geometric figure that has no boundaries, specifically a *torus*, and then measuring distances on this figure. Dacey stresses, however, that this is not a general solution in that it is unsuited to regions with irregular boundaries or to point patterns in which there is a pronounced directional trend. An application of toroidal mapping to a problem in point pattern analysis also is referenced in the following subsection.

Random Disturbance on a Uniform Pattern

The acceptance of the alternative hypothesis that a spatial pattern is nonrandom does not preclude the possibility that the pattern may have been generated by random disturbances of some other spatial pattern. Dacey and Tung (1962) analyzed this possibility in some detail. For any pattern of n points they postulated a relationship between a theoretical arrangement S and the observed arrangement P. The points in S have coordinates (x_i, y_i), whereas in P the actual locations are given as (u_i, v_i). A random disturbance exists when the actual locations are a function of the theoretical locations; that is,

$$(u_i, v_i) = f[(x_i, y_i), x, y],$$

where x and y are random variables, one for each coordinate. The requirement is that

$$[E(x)]^2 + [E(y)]^2 = R^2,$$

where R is "the mean random disturbance distance." If this disturbance factor is large compared to the area encompassed by S and the mean distance between points, then Dacey and Tung noted (1962, p. 84) that "all trace of the underlying theoretical pattern may be lost; inferentially, such a pattern appears to be generated by a random process." On the other hand, for a small value of R the theoretical pattern may still be discernible. Unfortunately, there are no tests available for deciding whether such a pattern "corresponds to the postulated pattern or should be treated as a random arrangement" (Dacey and Tung, 1962, p. 84).

In this section we discuss first the results of random disturbances on a uniform pattern. An observed pattern may appear to be in agreement with this "disturbance" model, but that does not necessarily mean that the pattern in question is uniform. It might after all have been generated by a random disturbance of a different magnitude on an alternative theoretical pattern.

Uniform patterns are discussed in many areas of location theory. The hexagonal location pattern of urban places in Christaller's model is perhaps the best known example. In a hexagonal arrangement of points the distance from a point to any one of the six other points surrounding it is $1.0750d^{-1/2}$, where d is the density of points per unit area.

In their discussion of random disturbances, Dacey and Tung used this expected distance for the hexagonal pattern to normalize the mean disturbance factor and thereby remove the effects of different densities and measurement scales in the theoretical pattern S. This normalized index D was given as

$$D = R/(1.0750d^{-1/2}).$$

For a number of selected values of D, random disturbances on a uniform pattern were generated. This generation involved for each point in the pattern, first, the random selection of a disturbance angle and, second, the random sampling of a value from a table of normal deviates. The absolute value of this deviate was scaled by multiplying it by D, and the resulting value gave the distance of disturbance. Hence, each point in the uniform pattern was disturbed in a randomly selected direction to a randomly chosen distance. For the generated patterns the mean nearest-neighbor distances for six sectors or regions were computed, and these distances were expressed then as a function of the density of points in the pattern. Table 5.6 contains the parameters which were derived in this way.

A set of these distance parameters was used by Dacey and Tung in testing the hypothesis that the central place system identified by Brush (1953) in southwestern Wisconsin (Figure 5.7) corresponds to a randomly disturbed hexagonal pattern with $D = 0.4$. The distance constants for this theoretical pattern are given in Table 5.6 and they are reproduced in the third column of Table 5.7. These values were compared with the observed constants derived from scaling the observed mean distances by the square root of the observed density. The "t" statistic proved not to be statistically significant, and the hypothesis above was accepted.

Dacey and Tung stressed that these results do not "imply that the $P_n(.4)$ provides the highest correspondence to the observed pattern," and that "a procedure is not given for obtaining that disturbance value D which is the best estimate of an observed pattern."

Dacey (1966c) has developed further the notions summarized in the first part of this subsection. The lattice of points in the hexagonal network was interpreted as an equilibrium state, and the displacements of the observed point locations from the equilibrium positions were considered as random variables. The object was to derive probability density functions for different spacing measures of the locational patterns. Specifically, interest was focused on the distance between an arbitrary point in the displaced set and the ith nearest point and the distance between a randomly located sample point or *locus* and the jth nearest point of the displaced set. The density functions for these distances were derived for two cases, one general in the sense that the distribution function for the displacement vector was not identified and the other more detailed in that an n-dimensional normal distribution was assumed for the displacement variable. For the latter case it was shown that "under mild restrictions the spacing between neighboring points is distributed as an order-statistic from the noncentral chi-square distribution."

The numerical evaluation of the integrals involved in the "normal" model was not possible, and a computer simulation was used to generate

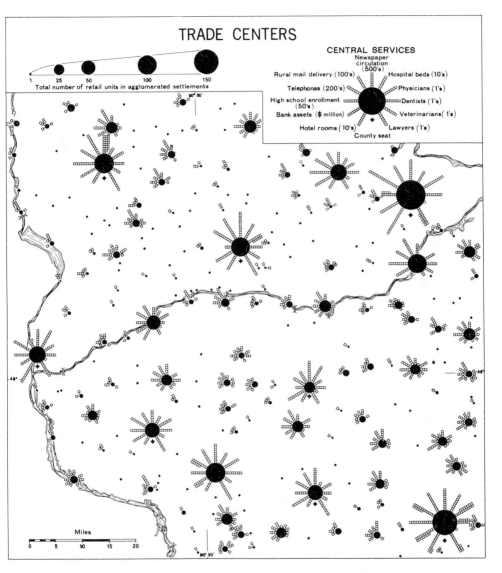

Figure 5.7. The central place hierarchy in southwestern Wisconsin. *Source:* Brush (1953).

estimates of point-to-point and locus-to-point order distances. To avoid operational difficulties associated with the possibility that some of the lattice points could be displaced into an area wider than that encompassed by the lattice itself, a mapping of the original hexagonal lattice onto a torus was employed. The boundary problems were reduced as a result.

TABLE 5.6.   REGIONAL NEIGHBOR DISTANCE PARAMETERS FOR
UNIFORM POINT PATTERNS CONTAINING VARYING
INTENSITIES OF RANDOM DISTURBANCE*

Re-gion k	Intensity index of random disturbance							
	0.0 (hexago-nal pattern)	0.05	0.2	0.4	0.6	0.8	1.0	Random pattern
1	1.0750	1.0142 (0.0046)	0.8143 (0.0177)	0.6056 (0.0344)	0.5240 (0.0334)	0.5190 (0.0379)	0.5035 (0.0356)	0.5000
2	1.0750	1.0373 (0.0039)	0.9220 (0.0150)	0.8941 (0.0302)	0.8199 (0.0315)	0.7686 (0.0353)	0.7708 (0.0362)	0.7863
3	1.0750	1.0603 (0.0034)	1.0635 (0.0128)	1.0732 (0.0227)	1.0520 (0.0279)	0.9797 (0.0299)	1.0620 (0.0362)	1.0403
4	1.0750	1.0834 (0.0031)	1.1853 (0.0199)	1.2594 (0.0170)	1.2906 (0.0325)	1.2875 (0.0414)	1.3067 (0.0465)	1.3034
5	1.0750	1.1064 (0.0045)	1.3497 (0.0255)	1.4779 (0.0288)	1.5453 (0.0401)	1.6044 (0.0532)	1.5345 (0.0426)	1.6175
6	1.0750	1.1295 (0.0049)	1.7656 (0.0147)	1.8238 (0.0429)	1.9025 (0.0451)	2.1075 (0.0613)	1.9070 (0.0467)	2.1010

*The expected distance to the nearest point in the kth of six regions is obtained by dividing the appropriate constant by $d^{1/2}$, where d is the density of points per unit area. The data within parentheses are sampling errors of the estimated constants.
Source: Dacey and Tung (1962, p. 90).

TABLE 5.7.   REGIONAL CONSTANTS FOR SOUTHWESTERN WISCONSIN

Region k	Regional constants as a function of density		t-Test of observed $P_u(0.4)$ constants	
	Central place pattern	$P_u(0.4)$	t value	$Pr(t)$
1	0.648	0.6056	1.2616	$0.2 < Pr(t) < 0.3$
2	0.858	0.8941	1.1954	$0.2 < Pr(t) < 0.3$
3	1.048	1.0732	1.1101	$0.2 < Pr(t) < 0.3$
4	1.258	1.2594	0.0824	$0.9 < Pr(t)$
5	1.449	1.4779	1.0035	$0.3 < Pr(t) < 0.4$
6	1.868	1.8239	1.0303	$0.3 < Pr(t) < 0.4$

Source: Dacey and Tung (1962, p. 95).

The "normal" model with a toroidal mapping of the points was tested empirically for a location pattern of 79 of the 99 largest urban places in Iowa. These 79 were located in the largest east-west rectangle that could be inscribed in the map of Iowa. The identification of this rectangle facilitated the toroidal mapping of the points. The correspondence between the spacing measures of the observed location pattern and those of alternative formulations of the "normal" model generally was quite high.

The conclusions drawn by Dacey have a wider philosophical significance than the immediate context might suggest. There is first a reiteration of the point made earlier in this text that in situations where there is no sampling error, then inferential tests are meaningless. Such was the case with the point-to-point distances in Dacey's study. A second and far more fundamental point is that even in the case of the sample locus-to-point distances there were no hypotheses or alternative hypotheses to evaluate. This shortcoming obviously stemmed from the fact that no substantive interpretations or theoretical justifications could be given to the displacement model. In pursuing this point with regard to possible interpretations of one of the model's parameters, Dacey was led to state his belief that "the most profitable line of research is to retain the notion of displacement from equilibrium positions but to incorporate these probabilistic notions within a more embracive statement of urban systems than allowed by the classic formulation of central place theory." The lesson should be clear—statistical analysis of data will be meaningful only when it is backed by theory, and, in turn, models and theories will be useful only when substantive interpretations can be given to their postulates and parameters.

Random Disturbance on a Multiple-clump Pattern

Dacey and Tung (1962) also consider the possibility of random disturbance on a multiple-clump pattern, defined in this case as follows: Clumps are uniformly distributed throughout the region with s points per clump and a constant density from clump to clump. Within each clump the points are in a compact circular arrangement around a central locus and at a distance ϵ from their respective nearest neighbors. In the generation of a random disturbance, each clump was assigned a central locus point. The same disturbance procedure outlined for the uniform pattern was then used to locate 27 points around this central locus in each clump. The regional neighbor distance parameters were derived then from the measurements made for these points in each clump. These parameters are tabled but are not included here.

Dacey (1966d) recently has sought to develop a model for a multiple-clump pattern that is of interest in urban geography. He considers the distribution of commercial districts throughout a large urban area and then the location of establishments within these districts or clumps. On the basis of certain assumptions, probability statements are derived for the expected distances between establishments in the same and different clumps.

Other Problems in Point Pattern Analysis

Dacey (1963) has considered two other topics related to point pattern analysis which can be handled by way of ordered neighbor distances and for which tests of randomness are known. The first concerns points distributed along radial lines as in a transportation network focused on a node. The points are assumed to be arranged on the radial network according to a Poisson distribution, and interest is focused on the expected distance from node *i*, "an end point of one or more radials," to the *j*th nearest point on any radial attached to *i*. This expected value is given as $E(r_{ij}) = j/cd$, where *j* relates to the order of neighbor, *c* is the average number of radials per node (of which there are *n*), and *d* is the mean number of points per unit length of radial. The test of randomness employs the chi-square statistic

$$\chi_j^2 = cd \sum_{i=1}^{n} r_{ij}$$

with *nj* degrees of freedom.

The second problem has to do with a pattern of lines. In this case, the description is "based upon the mean perpendicular distance from randomly selected points on lines of the pattern to each of the first, second, . . . , nearest lines" (Dacey 1963, p. 512). The test for randomness of the lines in *k*-dimensional space again involves a chi-square statistic.

Neither of these theoretical questions posed by Dacey has been considered in empirical geographic research, although line patterns particularly are of interest to the geographer.

Some of the geographer's reluctance in pursuing point pattern analysis may stem from an unwillingness to deal with abstractions of reality in the form of points. But such issues really evade the main question, for if the map area or space is defined carefully with respect to the "point" phenomenon under consideration, then useful results should be possible. The basic problem is how to interpret a finding of randomness or nonrandomness in the light of the processes which affect the phenomenon. These conceptual ties between substantive theorizing and pattern analysis demand careful consideration by geographers.

5.4   TESTS FOR RANDOMNESS IN THE DISTRIBUTION OF POINTS BY QUADRATS

The distances separating points do not summarize all the characteristics of a pattern. For example, the manner in which the points are distributed over a set of quadrats is another feature of a pattern which often is emphasized.

The possibility of testing for randomness in this distribution of points was implicit in our earlier discussions of particular probability laws and in the examples given of goodness-of-fit tests. It is now made explicit in the context of this section, but only by way of a brief review of the procedures which are involved.

In this *quadrat analysis*, as it is called, the testing of hypotheses as to the randomness of the distribution of points demands the comparison of the observed frequency distribution of quadrats with an expected one derived from a particular probability model. Those models which are applicable to purely random processes, and to more regular than random patterns in the case of both homogeneous and inhomogeneous regions, have been introduced already. In fitting these models, the moments of the observed distribution, for example, the observed density of points per quadrat, are used in estimating the parameters. Goodness-of-fit tests then are applied in the comparison of the observed and expected distributions. Dacey (1966b, 1967), Harvey (1966), and McConnell (1966a) illustrate the use of quadrat methods in geographic research. Chacko and Negi (1965) provide recent illustrations of the approach applied to problems in forest ecology.

The questions of quadrat size and shape are important considerations in this type of analysis. As the quadrats are made smaller compared to the total area, then the probability of a quadrat containing a point is decreased. Alternatively, if the quadrat size is increased, then the probability increases. The consequences of these two alternatives in regard to the estimation and interpretation of theoretical parameters have been discussed by Pielou (1957), Harvey (1966), and Dacey (1967). Some of these questions were touched upon in a discussion of the negative binomial model in Chapter 3. Curtiss and McIntosh (1950) suggest that the quadrat size should be $(2a/n)$, in other words, twice the mean area around each point in the pattern. This suggestion, however, was prompted in part by the desire for a fairly symmetric frequency distribution of the quadrats, which would better justify certain inferential tests involving quadrat statistics.

The question of quadrat shape is important only when some sampling of a point pattern is to be undertaken. In this context, the findings of Matern which were reviewed in Chapter 4 appear relevant. His conclusions that the differences in shape among such figures as a circle, regular hexagon, and square are insignificant in stratified random sampling are worth reiterating at this point.

5.5 RANDOMNESS IN THE ARRANGEMENT OF POINTS BY QUADRATS

Dacey (1966b) makes clear the distinction between a *distribution* and an *arrangement* of points among a set of quadrats or areal units by use of the simple diagram in Figure 5.8.

Distribution *A*

Value	Frequency
0	6
1	3
2	6

Arrangement *A*

0	0	1	2	2
0	0	1	2	2
0	0	1	2	2

Distribution *B*

Value	Frequency
0	6
1	3
2	6

Arrangement *B*

0	2	0	2	1
0	2	1	0	2
1	0	2	0	2

Figure 5.8. The frequency distributions *A* and *B* are identical, while the corresponding arrangements of values in cells or "counties" are clearly different. *Source:* Dacey (1966b).

The testing for randomness in the arrangement of points is not as straightforward as the corresponding problem for the distribution of points. The problem now is essentially one of determining whether the values representing the number of points per quadrat are randomly distributed with no significant tendencies towards a clustering of high and/or low values in particular groups of quadrats. There are a number of ways of handling this general problem, but in this section we concentrate upon only one of these, namely the use of *contiguity measures*.

As Dacey (1966b) points out, a random arrangement of values such as the number of points per quadrat can imply that there is at least no correlation among the values for adjacent quadrats. On this assumption it is possible to derive certain descriptions of a random arrangement of values in a two-dimensional space and to design tests of randomness based on these descriptions.

An original study on this question was published by Geary (1954). The problem as he stated it was "to determine whether statistics given for each county in a 'country' are distributed at random or whether they form a pattern." As a measure of this "patterning" or contiguity effect, Geary defined a contiguity ratio, *c*, as follows:

$$c = \frac{(n-1)}{2K_1} \cdot \frac{\sum'_{t \neq t'}(z_t - z'_t)^2}{\sum_t (z_t - \bar{z})^2},$$

where

n = total number of counties or areal units,

t = any one county,

Z = is the variable in question, for example, the number of points per quadrat,

$K_1 = \sum k_t$, where k_t is the number of connections or contiguous counties associated with county t,

$\sum_t$ = sum over all counties,

$\sum'$ = sum over contiguous counties.

The numerator of this expression involves the sum of the squared deviations between each county's Z value and the values in the contiguous counties. The denominator is the familiar sum of squared deviations from the mean. If the ratio is rewritten as

$$c = \frac{\frac{1}{2K_1} \cdot \sum'_{t \neq t'} (z_t - z'_t)^2}{\frac{1}{n-1} \sum_t (z_t - \bar{z})^2},$$

then, as Duncan (1961, p. 131) points out, it defines the ratio of within cluster variation to total variation. Assuming contiguity is high, then the ratio will be less than one. On the other hand, "when the statistics are distributed at random in the counties, the average value of the ratio is unity."

The theory concerning the contiguity ratio is developed along two lines by Geary. The first involves a randomization approach in which the observations comprise the universe and no assumption is made as to the nature of the underlying frequency distribution. The appropriate formulae for the mean and variance of c are derived. The second approach involves a discussion of the sampling theory of c, given that Z is normally distributed with mean zero and unknown variance. Then the first two moments of c are

$$\mu'_1 = 1$$
$$\mu'_2 = [n^2 k_1^2 + 2n(k_1 + k_2)][(n-1)/n^2(n+1)k_1^2],$$

where

n = number of observations,

$k_1 = K_1/n = \sum k_t/n$,

$k_2 = K_2/n = \sum k_t^2/n$.

The significance of the contiguity ratio then can be tested using the standard normal curve. The test statistic is

$$R = \frac{1-c}{\mu'_2}.$$

The null hypothesis is that the difference $(1 - c)$ is not statistically significant and that the distribution of values is random.

Dacey (1965b), in discussing this same problem, outlines a set of procedures which appear more operational than those suggested by Geary.

These procedures are a development of work initiated in the statistics literature on the problem of the arrangement and joins of colors in one, two,
three, and higher dimensions. Dacey (1965b) references much of this work,
as do Chacko and Negi (1965).

In the present context, however, we are interested only in Dacey's method
for describing a random arrangement of numerical values among quadrats.
Another form of contiguity problem which can be handled as a two- or
k-color map problem will be discussed in the following chapter on regression
analysis.

The discussion now follows Dacey (1965b, 1966b). Consider a set of
quadrats $(1, 2, \ldots, k, \ldots, m)$ for which there is measured a random variable
X, in this case the number of points per quadrat. We are interested in deriving
the expected number of pairs of quadrats for which one quadrat has i points
and the other j points, assuming that the arrangement of values is random.
The following terms are defined:

$L_k =$ number of quadrats adjacent to the kth. Quadrats are adjacent if they have an edge and/or a
vertex in common.

$$A = (1/2) \sum_{k=1}^{m} L_k,$$

$$D = \sum_{k=1}^{m} L_k (L_k - 1),$$

$$p(x) = P(X = x) \qquad \text{for } x = i, j.$$

Then the expected number of pairs which is sought is

$$E(i, j) = 2Ap(i)p(j), \qquad \text{for } i \neq j$$

and

$$E(i, i) = Ap(i)^2, \qquad \text{for } i = j.$$

The corresponding variance terms are

$$V(i, j) = 2Ap(i)p(j) + Dp(i)p(j)[p(i) + p(j)] \\ - 4(A + D)p^2(i)p^2(j), \qquad \text{for } i \neq j$$

and

$$V(i, i) = Ap^2(i) + Dp^3(i) - (A + D)p^4(i), \qquad \text{for } i = j.$$

In testing for randomness in an observed pattern of values, the ratio

$$R(i, j) = \frac{O(i, j) - E(i, j)}{[V(i, j)]^{1/2}},$$

where $O(i, j)$ is the observed number of pairs, is treated as a standardized
normal variate. The null hypothesis is that the arrangement is random.

An example of this test, taken from Dacey (1966b), is shown in Table

5.8. In this example, the test relates to the arrangement of the N largest urban places among the 99 counties of Iowa, and it is computed for six different values of N. Only one ratio value exceeds the critical value of $R(i,j)$ at the 95 percent confidence level; otherwise the hypothesis of a random arrangement is accepted in all cases.

TABLE 5.8.   ARRANGEMENT OF N LARGEST PLACES, IOWA, 1950

$R(i,j)$ ratios for arrangement

$N = 20$			$N = 40$		
	j			j	
	0	1		0	1	2
i 0	−0.12	−0.21	0	−1.04	1.50	−0.86
1		0.99	i 1		0.82	−0.84
			2			−0.29

$N = 60$			$N = 80$		
	j			j	
	0	1	2		0	1	2
0	−0.87	−0.08	−0.74	0	−0.38	−0.92	−0.40
i 1		1.25	0.04	i 1		0.54	1.21
2			−0.62	2			−0.37

$N = 100$				$N = 200$				
	j				j			
	0	1	2	3	1	2	3	4	5
0	−0.39	−0.80	−0.66	−0.14	1 0.08	−0.22	0.29	−1.32	0.15
1		0.72	0.21	0.18	2	0.26	1.23	−1.35	0.12
i 2			0.80	−0.81	i 3		1.26	−1.43	0.40
3				−0.22	4			0.06	0.67
					5				3.58

Source: Dacey (1966b, p. 538).

5.6 STATISTICAL CONSIDERATIONS IN CHOROPLETH PATTERNS

Some comments are made now on certain questions that arise in the statistical analysis of locational patterns of variables which are defined over fairly continuous areas and not merely at points. Climatic phenomena, soil and vegetation associations, crop patterns, and population density are but a few of the phenomena which are analyzed usually with reference to a system of areal units such as counties. These patterns, of course, are no different in character from the one discussed above in connection with analysis of the arrangement of points in a region. Instead of the number of points per quadrat, the random variable in question is now any one of a number of possibilities. The same problems of testing for randomness, and the use of the contiguity tests suggested by Geary and Dacey, apply equally well in these situations. In this section, however, we consider a different question in the description and analysis of these location patterns.

There are a number of location measures and indices which can be computed, for example, from county data. Isard (1960, Chapter 7) discusses many of these measures. In the present context, however, most of these are ignored simply because nothing is known about their sampling distributions, if indeed these are derivable in mathematical terms. Such is the case with the familiar location quotient and index of localization. Nevertheless there is one index introduced here for which some information is known concerning the sampling distribution and the associated moments. This is the Gini coefficient of concentration.

Consider a variable X measured over a set of areal units where the proportionate share of the total X associated with each unit is known. This spatial distribution is to be compared with another, a hypothetical distribution of X in which every areal unit has an equal share of the total.

The cumulative sums of the X values and of the number of areal units can be shown as the axes of a simple graph (Figure 5.9). On this graph the 45° line represents the curve of equal distribution or concentration. In other words, if every unit had an equal share of the total X, then the plot of the cumulative frequency distribution for X should coincide with this 45° line. The curve for the observed set of data, or indeed for sets of data for the same areal units, then can be plotted (Figure 5.9), and the Gini coefficient

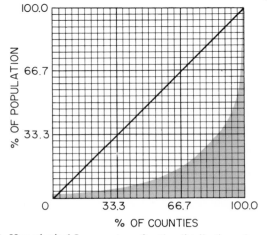

Figure 5.9. Hypothetical Lorenz curve for the distribution of population by counties.

computed as a measure of the departure of any one curve from the 45° line. The coefficient is computed as follows:

For a set of N observations, $x_1, x_2, \ldots, x_n$, define

$$d_{ij} = |x_i - x_j| \qquad \text{for } i \neq j,$$

where x_i and x_j are any two of the values. Then for the ith value alone define

$$d_i = \left[\frac{1}{N-1}\right] \sum_{j=1}^{N} |x_j - x_i| \qquad \text{for } i \neq j.$$

Further, the mean absolute deviation about a fixed x_i is given as

$$d = \frac{1}{N(N-1)} \sum_{i=1}^{N} \sum_{j=1}^{N} d \qquad \text{for } i \neq j$$

$$= \frac{1}{N} \sum_{i=1}^{N} d_i.$$

Consider, now, the case of a simple random sample n where $n \geqslant 2$, drawn without replacement. From the expressions above we have

$$d_i = \frac{1}{n-1} \sum_{j=1}^{n} |x_i - x_j| \qquad \text{for } i \neq j,$$

and

$$d = \frac{1}{n(n-1)} \sum_{i=1}^{n} \sum_{j=1}^{n} d_{ij} \qquad \text{for } i \neq j,$$

$$= \frac{1}{n} \sum_{i=1}^{n} d_i.$$

Glasser (1962) notes that if the sampling is done without replacement from an infinite population, and $n \geqslant 20$, then the sampling distribution of d is approximately normal. Then the estimated variance terms, based on the sample variance s^2, are

$$\hat{\sigma}^2_{dij} = 2 \cdot \frac{n}{n-1} (s^2 - d^2),$$

$$\hat{\sigma}^2_{di} = \frac{1}{n} \sum_{i=1}^{n} (d_i^2 - d^2),$$

and

$$\hat{\sigma}^2_d = \frac{2}{(n-2)(n-3)} [2(n-1)(\hat{\sigma}^2_{di}) - (\hat{\sigma}^2_{dij})].$$

The computation is facilitated by ranking the observations and defining

$$t_i = \sum_{j>i} x_j - \sum_{j<i} x_j$$

$$= t_{i-1} - x_{i-1} - x_i.$$

Then

$$d_i = \frac{1}{n-1} [t_i + (2_i - n - 1)x_i]$$

and

$$d = \frac{1}{n(n-1)} \sum_{i=1}^{n} (2_i - n - 1)x_i.$$

The Gini coefficient g is defined as

$$g = d/2\bar{x},$$

where $\bar{x}$ is the sample mean. The estimated variance of this statistic is approximated by the expression

$$\frac{1}{n\bar{x}^2} [\text{var}\,(d_i - gx_i)] \frac{N-n}{N}.$$

SUGGESTED READINGS

Bachi, R. (1963), "Standard Distance Measures and Related Methods for Spatial Analysis." *Papers*, The Regional Science Association, Vol. 10, pp. 83–132.

Dacey, M. F. (1960a), "A Note on the Derivation of Nearest-Neighbor Distances." *Journal of Regional Science*, Vol. 2, pp. 81–87.

Dacey, M. F. (1962), "Analysis of Central Place and Point Patterns by a Nearest-Neighbor Method." *Lund Studies in Geography, Series B*, Vol. 24, pp. 55–75.

Dacey, M. F. (1963), "Order Neighbor Statistics for a Class of Random Patterns in Multidimensional Space." *Annals*, Association of American Geographers, Vol. 53, pp. 505–515.

Geary, R. C. (1954), "The Contiguity Ratio and Statistical Mapping." *The Incorporated Statistician*, Vol. 5, pp. 115–145.

Getis, A. (1964), "Temporal Land-Use Patterns Analysis with the Use of Nearest-Neighbor and Quadrat Methods." *Annals*, Association of American Geographers, Vol. 54, pp. 391–399.

Greig-Smith, P. (1964), *Quantitative Plant Ecology*, Second edition. London: Butterworth & Co., Ltd., Chapter 3.

King, L. J. (1962a), "A Quantitative Expression of the Pattern of Urban Settlements in Selected Areas of the United States." *Tijdschrift voor Economische en Sociale Geografie*, Vol. 53, pp. 1–7.

Medvedkov, Y. V. (1964), "Applications of Mathematics to Some Problems in Economic Geography." *Soviet Geography*, Vol. 5, pp. 36–53.

Warntz, W. and Neft, D. (1960), "Contributions to a Statistical Methodology for Areal Distributions." *Journal of Regional Science*, Vol. 2, pp. 47–66.

ANALYSIS OF SPATIAL
RELATIONSHIPS AND AREAL
ASSOCIATIONS

A continuous theme in geographic research is that of analyzing the degree and direction of correspondence among two or more spatial patterns or locational arrangements. In some cases this analysis is guided by hypotheses about functional relationships or causal mechanisms between the phenomena in question; in other situations, the analysis is exploratory and seeks to derive inductive generalizations concerning the covariation of spatially distributed phenomena.

The early approach to this type of analysis was by way of map comparisons and, in some cases, map overlays. This chapter reviews the attempts to formalize this approach in the framework of regression and correlation analysis. In the final part of the chapter, emphasis is given to some of the technical problems which have become apparent as a result of these attempts.

6

6.1 SIMPLE REGRESSION AND
CORRELATION ANALYSIS IN GEOGRAPHY

A distinction between regression and correlation analysis does exist, but the two are so closely linked that it is seldom emphasized.

The regression model is appropriate for situations where a functional relationship is postulated. One variable then is considered to be dependent upon one or several others, being partly controlled by or related to them. Thus, the population density of a city census tract might be viewed as a function of the distance separating the tract from the center of the city. Similarly, the population size of a city at one point in time might be viewed as a function of its size in the previous time period.

By contrast, correlation analysis implies no dependence and considers only the covariation of the variables. The emphasis is on the degree to which the two sets of values vary together around their respective means and on the direction of this covariation. If it is positive, then the two sets of values increase or decrease together; if it is negative, then the one set increases as the other decreases.

The following section considers regression and correlation analysis in a simple two-variable situation.

The Linear Regression Model

Consider the data given in Table 6.1. For different values of X_1—the straight-line distance from the downtown of some hypothetical city—the population density X_0 has been computed for a selection of tracts. We are interested in studying the relationship between these two variables and in seeing how changes in the one variable, distance, affect the other.

A plotting of the observations in the diagram in Figure 6.1, which is called a *scatter diagram*, shows that for any particular value of X_1 there is a range of values for X_0. In general, however, the values of X_0 decrease as the value of X_1 increases. It is this average relationship that we wish to describe in terms of a mathematical function. The curve which shows the change in the average value of X_0 given X_1, for different values of X_1, is called a *regression curve*. It may take any one of a number of forms, but in this section only the linear regression model is considered.

The linear model may be written as

$$X_0 = \alpha + \beta X_1 + \epsilon,$$

where X_0 is the *dependent* variable, X_1 is the *independent* variable, α and β are parameters, and ϵ is a random error variate.

TABLE 6.1. POPULATION DENSITY AND DISTANCE FROM DOWN-
TOWN FOR SELECTED TRACTS IN A HYPOTHETICAL CITY

Tract i	Density (1000 persons per sq mi) X_0	Distance (1000 ft) X_1
1	27.5	5
2	21.0	5
3	14.5	5
4	20.8	10
5	20.0	10
6	16.6	10
7	11.0	10
8	19.1	15
9	17.2	15
10	10.7	15
11	7.8	15
12	15.2	20
13	11.6	20
14	9.8	20
15	13.2	25
16	10.6	25
17	7.9	25
18	6.5	25
19	9.0	30
20	6.7	30
21	6.2	30
22	3.9	30
23	7.9	35
24	5.3	35
25	1.9	35
26	5.3	40
27	3.4	40
28	3.0	40
29	1.7	45
30	1.0	45

Some comments are appropriate concerning these different terms. First, the distinction between a *dependent* and an *independent* variable is not always clear-cut, and it is possible that the same variable may appear in either role in different problems. In Figure 6.1 there are in fact two regression lines which can be plotted one representing the regression of X_0 on X_1 and the other the regression of X_1 on X_0. Only in the special case of a perfect relationship between the two variables will the two regression lines coincide. The problem of deciding which variable is the response or dependent variable and which one is the independent or control variable can be resolved only in the context

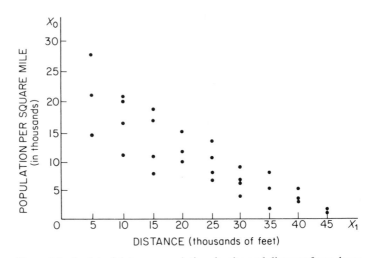

Figure 6.1. A plot of data on population density and distance from downtown for a sample of 30 census tracts in a hypothetical large city.

of a particular research situation. The same point holds true with regard to the choice of the particular model, linear or nonlinear, which is to be used. Hopefully, this choice will be suggested by the nature of the problem under investigation and by existing theory. The linear model is a restrictive one in the sense that it implies a constant rate of change in X_0 over the domain of X_1; nevertheless, it serves as a convenient approximation in many problem situations. Besides, we shall note later that often it is possible to apply transformations to data and thereby treat some nonlinear situation as linear ones.

The random error variate ϵ reminds us that we are dealing with a statistical relationship and not a precise, mathematical relationship. The contrast, in other words, is with a situation such as that represented in Figure 6.2. In this diagram the area of a square A is plotted against the length of the side of the square d. The relationship is an exact one, $A = d^2$, and there is no variance in this relationship. If indeed we were to measure the area with a planimeter and the length of the side with a ruler and plot these observed values in Figure 6.2, then the only reason for some of the observations not falling precisely on the curve would be measurement errors. Compare this situation with the one shown in Figure 6.1. Here there is a trend in the plotting of the

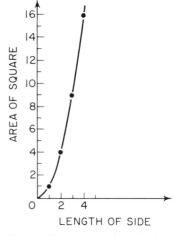

Figure 6.2. The relationship between area of square and the length of square side.

observations which is suggestive of a relationship, but obviously not an exact one, between the variables in question. There is some variance in the dependent variable for any particular value of X_1, and the regression model we noted describes only an average relationship. The deviations from this model are the terms ϵ_i, and these are assumed to be independent and normally distributed with mean zero and an unknown variance σ_ϵ^2. In applied research this assumption does not always hold true and there are attendent problems in the analysis which will be noted shortly.

The parameters α and β in the linear model have to be estimated from the sample evidence. The parameter α controls the elevation of the regression line; in other words, it gives the value of X_0 at which the regression line intersects the ordinate. The β value gives the slope of the regression line and it can be defined as the tangent of the angle which the regression line makes with the abscissa.

Estimation of the parameters gives the equation

$$Y_i = a + bX_{1i},$$

where Y_i is the predicted value of the dependent variable corresponding to X_{0i} and a and b are the estimates of α and β, respectively. The method of obtaining these estimates is the *least squares* technique, which provides estimates of α and β, namely a and b, such that the expression

$$U = \sum_{i=1}^{N} \epsilon_i^2 = \sum_{i=1}^{N} (X_{0i} - Y_i)^2 = \sum_{i=1}^{N} [X_{0i} - (a + bX_{1i})]^2$$

is minimized.

The least squares method is discussed in all standard books on statistics, including Hoel (1960), Fraser (1958), and Johnson and Leone (1964). Draper and Smith (1966) provide an excellent discussion of regression analysis in general.

The least squares procedure involves taking the partial derivatives of U with respect to a and b and setting these equal to zero. This derivation is given in Appendix A. As a result, two equations are obtained; these are referred to as the "normal equations":

$$Na + b \sum_{i=1}^{N} X_{1i} = \sum_{i=1}^{N} X_{0i}$$

and

$$a \sum_{i=1}^{N} X_{1i} + b \sum_{i=1}^{N} X_{1i}^2 = \sum_{i=1}^{N} X_{1i} X_{0i}.$$

These are solved to obtain the following expressions for a and b:

$$b = [\sum X_0 X_1 - (\sum X_0)(\sum X_1)/N]/[\sum X_1^2 - (\sum X_1)^2/N],$$
$$a = \bar{X}_0 - b\bar{X}_1.$$

It should be noted that for purposes of clarity the summation operators have been written without the identifying subscripts. Henceforth, unless otherwise stated, the summation will be understood to be over the values $i = 1, 2, 3, \ldots, N$.

The computation of these statistics for the data in Table 6.1 is shown in Table 6.2, and the corresponding regression line is plotted in Figure 6.3.

TABLE 6.2.   COMPUTATION OF REGRESSION COEFFICIENTS FOR DATA
IN TABLE 6.1.

$$N = 30$$

$$\sum X_0 = 316.30 \qquad\qquad \sum X_1 = 710.00$$
$$\sum X_0^2 = 4651.77 \qquad\qquad \sum X_1^2 = 21,200.00$$
$$\bar{X}_0 = 10.54 \qquad\qquad \bar{X}_1 = 23.67$$
$$\sum X_0 X_1 = 5400.00.$$

$$b = \frac{5400.00 - (316.30)(710.00)/30}{21,200.00 - (710.00)^2/30}$$
$$= (5400.00 - 7485.76)/(21,200.00 - 16,803.33)$$
$$= -2085.76/4396.67$$
$$= -0.47.$$
$$a = 10.54 - (-0.47)(23.67)$$
$$= 10.54 + 11.12$$
$$= 21.66.$$

It is important to remember that when a sample is involved, the statistics a and b are estimators of the population parameters and are derived on the basis of certain assumptions, the main ones being as follows:

(i)   The regression analysis assumes that the independent variable

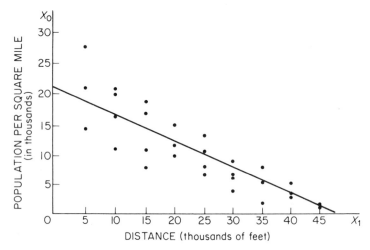

Figure 6.3. Linear regression curve fitted to data in Figure 6.1. The parameters for this curve are given in Table 6.2.

is not subject to error and may even be controlled. In geographic research the latter case is seldom encountered and the former situation seldom emphasized. While this assumption imposes no restraints on the solution, it does influence the interpretations which can be drawn from the analysis. If the independent variable also is subject to error, then the linear regression model described here may be inappropriate for describing the relationship.

(ii)   As noted earlier, the deviations $(X_{0i} - Y_i)$ are assumed to be independent and normally distributed. In geographic analyses involving county or quadrat data, this assumption often is a troublesome one and is not always satisfied. Questions of spatial autocorrelation become significant in this context, and these will be dealt with later in this chapter.

(iii)   Formally, the regression model assumes that for each value of X_1 there is a conditional distribution of X_0 values. In actual research problems, usually there is only one X_0 measurement for each X_1 value, but this does not invalidate the theory which assumes that these conditional distributions are equally variable. This is the assumption of *homoscedacity*, and unless it is satisfied, the least squares solution does not yield the best estimates of the parameters.

These assumptions appear to be of varying importance, depending on whether or not one is working in a formal inferential framework. When this is not the case, as in much geographic research, the problems stemming from spatial autocorrelation are perhaps the most formidable.

In a regression analysis it is of interest to know how closely or precisely the values of the dependent variable can be estimated from the values of the independent variable. The answer to this question involves consideration of some different sums of squares terms.

Examine the situation represented in Figure 6.4 for, say, the *i*th observation. The deviation of this observation from the mean value for the dependent variable is represented by the line segment *ab*. Over all *N* observations, this sum of squares for total variation is given by the familiar expression $\sum_i (X_{0i} - \bar{X}_0)^2$. For the same *i*th observation, the portion of the total variation which is accounted for by the regression on X_1 is the line segment *cb*, which represents the deviation between Y and $\bar{X}_0$. Again, when summed over all *N* observations, this yields the sum of squares associated with regression, or the regression sum of squares, $\sum_i (Y_i - \bar{X}_0)^2$. The portions of the total variation which remain "unexplained" or unaccounted for are the line segment *ac* in the case of the *i*th observation and the expression $\sum_i (X_{0i} - Y_i)^2$ for all *N* observations.

The discussion above is summarized as follows:

$$\begin{pmatrix} \text{Total sums of} \\ \text{squares} \\ (ss_{\text{tot}}) \end{pmatrix} = \begin{pmatrix} \text{regression sums} \\ \text{of squares} \\ (ss_{\text{reg}}) \end{pmatrix} + \begin{pmatrix} \text{residual sums} \\ \text{of squares} \\ (ss_{\text{resid}}) \end{pmatrix}.$$

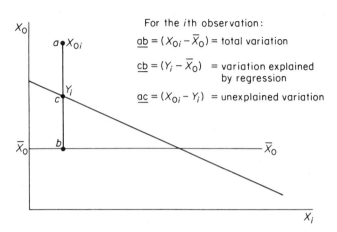

Figure 6.4. Diagrammatic representation of the different sources of variation in the simple linear regression problem.

The computation of these sums of squares expressions is facilitated by use of the following formulae:

$$ss_{\text{tot}} = \sum X_0^2 - (\sum X_0)^2/N,$$

$$ss_{\text{reg}} = b[\sum X_0 X_1 - (\sum X_0)(\sum X_1)/N]$$
$$= a \sum X_0 + b \sum X_0 X_1 - (\sum X_0)^2/N,$$

$$ss_{\text{resid}} = ss_{\text{tot}} - ss_{\text{reg}}$$
$$= \sum X_0^2 - a \sum X_0 - b \sum X_0 X_1.$$

The corresponding computations for the data in Table 6.1 are summarized below.

$$ss_{\text{tot}} = 4651.77 - (316.30)^2/30$$
$$= 4651.77 - 3334.80$$
$$= 1316.97.$$

$$ss_{\text{reg}} = -0.47[5400.00 - (316.30)(710.00)/30]$$
$$= -0.47(5400.00 - 7485.76)$$
$$= -0.47(-2085.76)$$
$$= 980.31.$$

$$ss_{\text{resid}} = ss_{\text{tot}} - ss_{\text{reg}}$$
$$= 1316.97 - 980.31$$
$$= 336.66.$$

We might note that if the ss_{resid} had been computed independently with the formula given above, then rounding error would have accounted for some discrepancy in the results. Computational accuracy always is increased with more decimal places being carried.

The residual sums of squares allows us to estimate the variance about the regression line, which is $\sigma^2_{0.1}$. The subscripts in this case identify the particular random variables X which are involved in the regression. The expression for this estimate is

$$s^2 = ss_{\text{resid}}/(N - 2).$$

The square root of this expression is known as the *standard error of estimate*.

Draper and Smith (1966, p. 17) point out that $\sigma^2_{0.1}$ may or may not be equal to σ^2_ϵ, which we mentioned earlier was the unknown variance of the normal variate ϵ. If the two values are equal, then "the postulated model is the true model" and s^2 in this case is also a correct estimate of σ^2_ϵ. On the other hand, if $\sigma^2_{0.1} > \sigma^2_\epsilon$, then the postulated model is incorrect.

Usually it is desirable that the standard error of estimate be relatively small, at least in comparison to the observed mean value of the dependent variable.

Questions of statistical inference are relevant with respect to the parameters α and β. When sampling is involved, the estimators a and b are of course random variables for which there are sampling distributions. It is possible, therefore, to derive confidence intervals for α, β, and $\sigma^2_{0.1}$. Since these questions have not been emphasized in the geography literature, the interested reader is referred to one of the statistical texts cited earlier for a discussion of the appropriate procedures.

In some geographic studies, for example, Thomas (1961) and Olsson (1965), the hypothesis that $\beta = 0$ is tested. The test statistic in this case is

$$t = b/(s_{0.1}/[\sum (X_1 - \bar{X}_1)^2]^{1/2}),$$

with $(N - 2)$ degrees of freedom. The part of this expression to the right of the first division slash gives the estimated standard error of b. Acceptance of the null hypothesis prompts the conclusion that in the population from which the sample is drawn there is no significant regression of X_0 on X_1. Again, these results appear meaningless if no sampling is involved.

There are a number of other tests of significance which apply in regression analysis. The hypotheses that $\alpha = 0$ and $\sigma^2_{0.1} = 0$ are two such examples. The significance of the difference between two regression coefficients also can be tested, a rejection of the null hypothesis in this case suggesting that the two regression lines are not parallel.

There are several examples in the geography literature of simple regression analysis. Two will suffice here. Haggett (1964), in his analysis of the distribution of forest in southeast Brazil, fitted linear regression models to a set of 24 observations for selected independent variables that were hypothesized as important in explaining the proportion of area forested (X_0). The equations were as follows:

(i) $X_0 = \log 3.39 + 0.31 \log X_1$,   where X_1 is a terrain index.

(ii) $X_0 = \log 24.6 - 0.44 \log X_2$,   where X_2 is a settlement spacing index.

(iii) $X_0 = \log 57.3 - 0.25 \log X_3$,   where X_3 is rural population density.

(iv) $X_0 = \log 0.02 + 1.38 \log X_4$,   where X_4 is a forest density index.

(v) $X_0 = \log 34.7 - 0.21 \log X_5$,   where X_5 is land value.

No significance tests were published in this study, but the signs of the regression coefficients were consistent with the hypothesized directions of the relationships.

Olsson (1965), in studying the relationships between migration streams and the sizes of urban places, employed a gravity model in the form of a regression analysis. The gravity model is well-known in geographic research and is discussed in Lukermann and Porter (1960). Olsson notes that it is expressed usually as

$$I_{ij} = k \cdot P_i \cdot P_j / D_{ij}^b,$$

where I_{ij} is the level of interaction (in Olsson's study, migration) between places i and j, P_i and P_j are the sizes of the places i and j, D_{ij} is the distance between i and j, and k and b are empirically derived constants. To estimate these constants, Olsson modified the gravity model somewhat by regressing (I_{ij}/P_iP_j) as the dependent variable on the distance variable D_{ij}. Thus he worked with the linear regression equation

$$\log (I_{ij}/P_iP_j) = \log k - b \log D_{ij}$$

and obtained the sets of regression coefficients given in Table 6.3. These relate only to migration to centers larger than the place of origin; the values for migrations to smaller places are not reproduced here. The results in Table 6.3 prompted the following conclusions by Olsson concerning the study hypotheses:

(i)   The fact that the absolute value of b (that is, ignoring the sign) for category F (total migration to all size groups) decreased as the population size of the origin increased was consistent with the hypothesis that "migrants from small places move shorter distances than migrants from large places." In other words, the migration intensity associated with smaller origins appeared more sensitive to changes in distance. Olsson noted, however, that some of the differences between these b values were not significant statistically, as shown in Table 6.4. A pooling of the groups B, C, and D did produce significant differences between the b values for this category F.

(ii)   The b values (again in absolute terms) decreased numerically in a manner consistent with the hypothesis that "migration distances decrease both with the size of the place of outmigration and with the size of the destination." Compare, for example, the following values:

TABLE 6.3.   SUMMARY TABLE FOR MIGRATIONS TO LARGER PLACES

		To hierarchical group															
From	A			B			C			D			E			F		
	k	b	r	k	b	r	k	b	r	k	b	r	k	b	r	k	b	r
A	1.88	−1.81*	−.95*	1.58	−1.54*	−.89*	1.69	−1.49†	−.97*	1.64	−1.54*	−.96*	1.16	−1.01*	−.85*	1.89	−1.92*	−.95*
B				1.47	−1.10*	−.90*	1.58	−1.25†	−.94*	1.44	−1.16*	−.95*	.92	−.66*	−.82*	1.16	−1.59*	−.94*
C							1.58	−1.38*	−.89*	1.60	−1.30*	−.87*	1.05	−.78*	−.74*	1.65	−1.56*	−.91*
D										1.77	−1.29*	−.87*	1.77	−1.40*	−.83*	1.61	−1.49*	−.87*
E													.53	+.29†	+.10†	1.04	−1.10*	−.74*

*Significant on the 99 percent level. †Not Significant. k = intercept with the y axis. b = slope of gravity line. r = simple correlation coefficient.

Source: Olsson (1965, p. 29).

$$b_{AA} = 1.81, \quad b_{AB} = -1.54 \ldots b_{AE} = -1.01$$
$$b_{BB} = -1.10 \ldots b_{EE} = -0.66$$
$$\ldots$$
$$\ldots$$
$$b_{EE} = +0.29$$

Again not all the differences were statistically significant (Table 6.5), although they were when groups *B*, *C*, and *D* again were pooled.

Olsson's study illustrates the use of regression analysis in the testing of substantive hypotheses. Also, there is the use of inferential tests concerning the significance of differences between statistics to qualify and reinforce the interpretations.

The use of logarithms in both of the examples discussed above is illustrative of a frequently used strategy in regression analysis. In many cases, the relationship between variables in their original form is nonlinear, but between transformations of these variables it is linear. For example, the nonlinear equation $X_0 = e^{\alpha + \beta X_1}$ can be written as a linear one in the logarithms of X_0 and X_1 as follows:

$$\ln X_0 = \alpha + \beta X_1.$$

Olsson's treatment of the gravity model was along the same lines. Similarly, the discussion by Berry, et al. (1963) of the relationship between population density and distance from the center of the city suggests that the model which is appropriate is

$$d_x = d_0 e^{-bx},$$

where "d_x is the population density d at distance x from the city center, d_0 is central density as extrapolated, and b is the density gradient indicating the rate of diminution of density with distance, a negative exponential decline." We referred to this model earlier regarding the probability law it suggests. Now we consider it as a regression model. When the model is written in natural logarithms, it becomes

$$\ln d_x = \ln d_0 - bX,$$

which is linear and can be fitted by least squares methods.

Again, the model $P = A(B)^N$, which Berry and Garrison (1958) used to describe the relationship between the population size of central places (P) and the number of businesses for any function (N), is linear in logarithms. Thus,

$$\log P = \log A + N \log B.$$

In the Berry and Garrison study, the antilogarithms for A and B were obtained, and for $N = 1$ these two values were multiplied to yield estimates of the "threshold population" for each function.

TABLE 6.4. SUMMARY OF SIGNIFICANCE TESTS OF DIFFERENCES IN b VALUES FOR TOTAL OUT-MIGRATION FROM EACH SIZE GROUP

Pair of b values	Level of significance, percent			
	99	95	90	90
AF^b–BF^b				
CF^b		X		
DF^b		X		
EF^b	X			
BF^b–CF^b				X
DF^b				X
EF^b		X		
CF^b–DF^b				X
EF^b		X		
DF^b–EF^b			X	

Source: Olsson (1965, p. 30).

TABLE 6.5. SUMMARY OF SIGNIFICANCE TESTS OF b VALUES FOR MIGRATIONS TO LARGER PLACES

Pair of b values	Level of significance, percent			
	99	95	90	90
AA^b–AB^b				X
AC^b		X		
AD^b		X		
AE^b	X			
AB^b–AC^b				X
AD^b				X
AE^b		X		
AC^b–AD^b	X			X
AE^b	X			
AD^b–AE^b				X
BB^b–BC^b		X		X
BD^b	X			X
BE^b				
BC^b–BD^b	X			X
BE^b				
BD^b–BE^b				X
CC^b–CD^b	X			X
CE^b				
CD^b–CE^b		X		
DD^b–DE^b				X

Source: Olsson (1965, p. 30).

Correlation Analysis

For a sample of N observations and paired values of two random variable X_0 and X_1, the *correlation coefficient* is defined as

$$r_{01} = [\sum X_0 X_1 - (\sum X_0)(\sum X_1)/N]/[\sum X_0^2 - (\sum X_0)^2/N]^{1/2}$$
$$\cdot [\sum X_1^2 - (\sum X_1)^2/N]^{1/2}.$$

The distinction between dependent and independent variables is unimportant in the expression above. The formal theory in this case assumes simply a random sample from a bivariate normal population. The best estimator of the correlation coefficient ρ for this population is the sample correlation coefficient r, defined above. This sample coefficient can be used in deriving confidence intervals for ρ and in testing hypotheses concerning this parameter. Again, the most commonly tested hypothesis is that $\rho = 0$. The critical values of r for given confidence levels in testing this hypothesis are available in published tables.

In geographic research to date, correlation analysis has appeared usually in a more informal role. The point already has been made about the frequent absence of formal sampling designs in geographic research and the fact that most inferential questions appear meaningless as a result. Nevertheless, it often is possible to gain significant information from the correlation analysis irrespective of the inferential questions which may be ignored. The connections between correlation and regression analyses are important in this regard.

First, the ratio of the regression sum of squares to the total sum of squares defines the *coefficient of determination r^2*, which is the square of the correlation coefficient. It is in this sense that the phrase "level of explained variation" is used when discussing the relationship between dependent and independent variables. A certain proportion of the variation in the dependent variable is attributable to, or "explained by," covariation with the independent variable. Whether there is any causal mechanism linking the two variables or whether one is conditional upon the other are broader but nonetheless important questions which *never* are resolved by the statistical analysis alone.

Second, it is worth noting that the correlation coefficient r and the regression coefficient b are directly related. Recall that

$$b = [\sum X_0 X_1 - (\sum X_0)(\sum X_1)/N]/[\sum X_1^2 - (\sum X_1)^2/N].$$

The denominator of this expression obviously can be written as

$$[\sum X_1^2 - (\sum X_1)^2/N]^{1/2} \cdot [\sum X_1^2 - (\sum X_1)^2/N]^{1/2}.$$

Now if we adopt this last convention, and if we multiply both the numerator and denominator of the expression for b by $[\sum X_0^2 - (\sum X_0)^2/N]^{1/2}$, we obtain

$$b = \frac{[\sum X_0 X_1 - (\sum X_0)(\sum X_1)/N] \cdot [\sum X_0^2 - (\sum X_0)^2/N]^{1/2}}{[\sum X_1^2 - (\sum X_1)^2/N]^{1/2} \cdot [\sum X_0^2 - (\sum X_0)^2/N]^{1/2} \cdot [\sum X_1^2 - (\sum X_1)^2/N]^{1/2}}$$

$$= r_{01} \cdot \frac{[\sum X_0^2 - (\sum X_0)^2/N]^{1/2}}{[\sum X_1^2 - (\sum X_1)^2/N]^{1/2}}$$

$$= r_{01} \cdot \frac{(N-1)^{1/2} s_0}{(N-1)^{1/2} s_1}$$

$$= r_{01} \cdot \frac{s_0}{s_1}.$$

In other words, the regression coefficient is simply the correlation coefficient scaled by the ratio of the standard deviation for the dependent variable to the standard deviation for the independent variable. The two coefficients obviously always have the same sign.

In the studies by Haggett and Olsson referred to earlier, the correlation coefficients were computed. Those for Olsson's study were included in Table 6.3. The coefficients of determination for Haggett's study are given in Table 6.6. The values in this case have been converted into percentage figures.

TABLE 6.6.   SUDESTE: RESULTS OF REGRESSION ANALYSIS OF FACTOR EFFECTS ON FOREST COVER

Factors	Transformation	Relationship	Variance reduction (percent)
Terrain index (X_1)	logarithmic	direct	38.4
Settlement spacing index (X_2)	logarithmic	direct	7.2
Rural population density index (X_3)	logarithmic	inverse	19.4
Forest density index (X_4)	logarithmic	direct	47.5
Land values index (X_5)	logarithmic	inverse	13.2

Source: Haggett (1964, p. 374).

In a study of the relationships between urban population size and distance to the nearest neighboring urban center of the same size in Iowa, Thomas (1962) illustrated the use of inferential procedures in testing the significance of the difference between correlation values. The appropriate test demands that a simple transformation be made. The correlation coefficient has a non-symmetrical sampling distribution, and it is convenient to transform the coefficient into a z statistic which is normally distributed. The transformation is of the form

$$z = \frac{1}{2} \ln \frac{1+r}{1-r},$$

where "ln" denotes the natural logarithm. Then to test the null hypothesis H_0: $\rho_1 = \rho_2$, we use the statistic

$$u = (z_1 - z_2)\Big/\Big(\frac{1}{n_1 - 3} + \frac{1}{n_2 - 3}\Big)^{1/2}$$

which is approximately normally distributed with mean zero and variance one.

TABLE 6.7. STABILITY OF DISTANCE–POPULATION SIZE RELATIONSHIPS

Correlation Coefficients and z Transformations for the Distance Between a Sample City and Its Nearest Neighbor of the Same Population Size and the Population Size of the Sample City: Hypothesis I.

	1900	1910	1920	1930	1940	1950
Correlation coefficients	0.30*	0.64*	0.57*	0.57*	0.65*	0.59*
Z transformations	0.3095	0.7582	0.6475	0.6475	0.7753	0.6777

*Significantly large at the 95 percent confidence level.

The Significance of the Difference Between Correlation Coefficients for Each Date: Hypothesis I.

Difference in values	1910	1920	1930	1940	1950
1900	0.4487*	0.3380*	0.3380*	0.4658*	0.3682*
1910		0.1107	0.1107	0.0029	0.0805
1920			0.0000	0.1278	0.0302
1930				0.1278	0.0302
1940					0.0492

*Significantly large at the 95 percent confidence level.
Source: Thomas (1962, p. 19).

The results which Thomas obtained (Table 6.7) prompted the conclusion that if 1900 is ignored, then there has been remarkable stability in the size-distance relationships, notwithstanding the social and economic changes that have taken place over the period.

Whereas Thomas was concerned with correlation values for different time periods, the same type of test can be used in comparing correlation coefficients for different regions. The test, in this case, either supports or contradicts the notion that particular variables are correlated to the same degree in the various regions.

The computation of the correlation coefficient for the data in Table 6.1 is outlined below.

$$r = \frac{5400.00 - (316.30)(710.00)/30}{[4651.77 - 3334.80]^{1/2}[21,200.00 - 16,803.33]^{1/2}}$$

$$= \frac{5400.00 - 7485.76}{[(1316.97)(4396.67)]^{1/2}}$$

$$= -2085.76/(5,790,282.49)^{1/2}$$

$$= -2085.76/2406.30$$

$$= -0.89$$

$$r^2 = 0.79.$$

6.2 THE CORRESPONDENCE OF POINT PATTERNS

The examples of regression analysis cited up to this point have been essentially one-dimensional problems in that the values of the dependent variable can be represented as points distributed on a linear scale. As we have noted earlier, however, two-dimensional point patterns are of great interest in geography, and questions concerning the correspondence of different point patterns often are raised. Tobler (1965) has discussed this problem of comparing two-dimensional distributions and his analysis is summarized here.

Consider a set of N paired observations W_j and Z_j, where W_j is a location in Cartesian space (u_j, v_j) and Z_j is similarly defined as a location with coordinates (x_j, y_j). An example would be the home location in a region of a migrant at one point in time and his location in the same region at a later point in time.

Tobler interprets the symbols W_j and Z_j in two ways. First, they are regarded as complex numbers. Thus,

$$W_j = (u_j + iv_j)$$

and

$$Z_j = (x_j + iy_j),$$

where, algebraically, i obeys the rule $i^2 = -1$.

The subject of complex numbers is considered in most introductory mathematical texts. Rudin (1964), for instance, presents a discussion of the topic at this level. Allen (1959) and Frisch (1966) in their discussions of mathematical economics also outline the essentials of the subject of complex numbers.

It suffices to note here that in the two-dimensional regression problem which requires analysis of points represented in rectangular coordinates the algebra of complex numbers affords definite advantages in deriving a mathematical solution. The points are handled as complex numbers which can be added and multiplied according to some clearly stated rules. For example,

$$W_j + Z_j = (u_j + iv_j) + (x_j + iy_j)$$
$$= (u_j + x_j) + i(v_j + y_j),$$

and

$$W_jZ_j = (u_j + iv_j) \cdot (x_j + iy_j)$$
$$= u_jx_j + u_jiy_j + iv_jx_j + i^2v_jy_j$$
$$= u_jx_j + iu_jy_j + iv_jx_j - v_jy_j \qquad \text{for } i^2 = -1.$$
$$= (u_jx_j - v_jy_j) + i(u_jy_j + v_jx_j).$$

Tobler notes that least squares methods can be used to estimate the coefficients A and B in the transformation

$$\hat{W}_j = A + BZ_j,$$

where $A = (a_1 + ia_2)$ and $B = (b_1 + ib_2)$. The expression which is minimized in this case is

$$\sum_{j=1}^{N} |\hat{W}_j - W_j|^2$$

In terms of the rectangular coordinates the complete transformation equations are

$$\hat{u}_j = a_1 + b_1x_j - b_2y_j,$$
$$\hat{v}_j = a_2 + b_2x_j + b_1y_j.$$

This situation is shown diagrammatically in Figure 6.5, although for convenience in this case both points for the jth observation are shown in the same plane.

A coefficient of determination is defined in this context as follows:

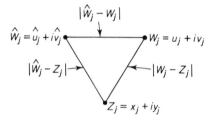

Figure 6.5. Two-dimensional regression problem. Here the two points W_j and Z_j are shown in the same plane; the Z_j point pattern is the "independent variable" and the W_j pattern is the "dependent." *Source:* Tobler (unpublished).

$$R^2_{wz} = \sum |\hat{W} - Z|^2 / \sum |W - Z|^2$$
$$= 1 - (\sum |\hat{W} - W|^2 / \sum |W - Z|^2).$$

The alternative approach to the two-dimensional problem which Tobler discussed interprets W_j and Z_j as vectors. That is,

$$W_j = [u_j, v_j]^T$$

and

$$Z_j = [x_j, y_j]^T.$$

Then the constants are

$$A = (a_1, a_2)^T$$

and

$$B = \begin{bmatrix} b_{11} & b_{12} \\ b_{21} & b_{22} \end{bmatrix}.$$

The complete transformation becomes

$$\begin{bmatrix} \hat{u}_j \\ \hat{v}_j \end{bmatrix} = \begin{bmatrix} a_1 \\ a_2 \end{bmatrix} + \begin{bmatrix} b_{11} & b_{12} \\ b_{21} & b_{22} \end{bmatrix} \cdot \begin{bmatrix} x_j \\ y_j \end{bmatrix}.$$

Again, as Tobler notes, a correlation coefficient can be defined but it involves a lengthy expression which is not included here.

This type of two-dimensional analysis has been used in the physical sciences but it has yet to be developed in social science research. In line with the stimulus-response interpretation given to ordinary regression analysis in the context of much agricultural research, Tobler (1965, p. 138) suggested that a "two-dimensional interpretation might be that a change in one geographical pattern is followed by a change in a second geographical pattern." As he stresses with regard to an empirical example concerning the correspondence between two location patterns of the residences of marriage partners, however, the theoretical underpinnings of such an analysis may be very weak. As a consequence, "the analysis must remain statistical and devoid of fruitful substantive interpretation." Again then, we have a reiteration of the point which has been made often in this text, that statistical analysis appears most powerful when it is backed by substantive theory and hypotheses.

6.3 MULTIPLE REGRESSION AND CORRELATION ANALYSIS

The majority of spatial distributions in which geographers are interested are typically so complex in their structure and relationships that they cannot be explained satisfactorily in terms of only one independent variable. The use of regression and correlation models with two or more independent variables in these situations involves no great differences in model conceptualization from that of the simple two-variable case. The computational problems, however, are increased considerably.

We continue to discuss only the linear model. The example of two independent variables will be emphasized, although the generalization to more than two such variables will be noted.

Solution of Normal Equations

Whereas in the case of one independent variable we were concerned with the fitting of a straight line to our observations, now we are interested in the fitting of a linear surface such that the sum of the squared deviations from this surface is minimized. As before, we seek to minimize the expression

$$U = \sum (X_{0i} - Y_i)^2,$$

only now in the case of two independent variables it is equal to

$$\sum [X_{0i} - (a + b_1 X_{1i} + b_2 X_{2i})]^2.$$

There are three parameters to be estimated and the least squares technique is employed. The partial derivatives of U with respect to b_1, b_2, and a are set equal to zero, and the following three normal equations are obtained:

$$\sum X_{0i} X_{1i} = b_1 \sum X_{1i}^2 + b_2 \sum X_{1i} X_{2i} + a \sum X_{1i}$$

$$\sum X_{0i} X_{2i} = b_1 \sum X_{1i} X_{2i} + b_2 \sum X_{2i}^2 + a \sum X_{2i}$$

$$\sum X_{0i} = b_1 \sum X_{1i} + b_2 \sum X_{2i} + Na.$$

As they appear here, these equations could be solved simultaneously once the necessary sums of squares and cross products have been obtained. When the model is generalized for more than two independent variables, however, this solution appears more difficult. In the case of m variables, of which $(m - 1)$ are independent, the set of normal equations will involve $(m - 1)$ expressions of the form

$$\sum X_{0i} X_{ji} = b_1 \sum X_{1i} X_{ji} + b_2 \sum X_{2i} X_{ji} + \ldots + b_j \sum X_{ji}^2 + \ldots + a \sum X_{ji},$$

plus the equation

$$\sum X_{0i} = b_1 \sum X_{1i} + b_2 \sum X_{2i} + \ldots + b_j \sum X_{ji} + \ldots + Na.$$

If m is as large as even 5 or 6, then the solution of these equations requires the use of a standardized procedure. We shall discuss the use of matrix algebra in this context. Some other well-known procedures are reviewed in Ezekiel and Fox (1959) and Johnson and Leone (1964). However, the formulation in terms of matrices is the one used in computer-derived solutions to the problem; hence it is favored in most recent books on the subject, for example, Draper and Smith (1966).

A computational shortcut involves rewriting the normal equations in deviation form. Recall that

$$\sum x_{0i}^2 = \sum (X_{0i} - \bar{X}_0)^2 = \sum X_{0i}^2 - (\sum X_{0i})^2/N,$$

and

$$\sum x_{1i}^2 = \sum (X_{1i} - \bar{X}_1)^2 = \sum X_{1i}^2 - (\sum X_{1i})^2/N.$$

A similar expression can be derived for $\sum x_{2i}^2$. The cross products terms are written as follows:

$$\sum x_{0i} x_{1i} = \sum (X_{0i} - \bar{X}_0)(X_{1i} - \bar{X}_1) = \sum X_{0i} X_{1i} - (\sum X_{0i})(\sum X_{1i})/N,$$

$$\sum x_{0i} x_{2i} = \sum (X_{0i} - \bar{X}_0)(X_{2i} - \bar{X}_2) = \sum X_{0i} X_{2i} - (\sum X_{0i})(\sum X_{2i})/N,$$

$$\sum x_{1i} x_{2i} = \sum (X_{1i} - \bar{X}_1)(X_{2i} - \bar{X}_2) = \sum X_{1i} X_{2i} - (\sum X_{1i})(\sum X_{2i})/N.$$

By substituting these expressions in the normal equations, we obtain

$$\sum x_{0i} x_{1i} = b_1 \sum x_{1i}^2 + b_2 \sum x_{1i} x_{2i},$$

and

$$\sum x_{0i}x_{2i} = b_1 \sum x_{1i}x_{2i} + b_2 \sum x_{2i}^2.$$

The third equation which we had above is no longer required.

For the generalized case of $(m - 1)$ independent variables, the normal equations in deviation form are given by

$$\sum x_{0i}x_{ji} = b_1 \sum x_{1i}x_{ji} + b_2 \sum x_{2i}x_{ji} + \ldots + b_j \sum x_{ji}^2,$$
$$j = 1, 2, \ldots, (m - 1).$$

The solution for the parameters remains formidable, even with the equations in deviation form. At this point, then, we make use of the matrix algebra which is reviewed in Appendix A.

Consider the two normal equations in deviation form given above. A knowledge of simple matrix manipulation allows these equations to be written in the following form:

$$\begin{bmatrix} \sum x_0 x_1 \\ \sum x_0 x_2 \end{bmatrix} = \begin{bmatrix} \sum x_1^2 & \sum x_1 x_2 \\ \sum x_1 x_2 & \sum x_2^2 \end{bmatrix} \cdot \begin{bmatrix} b_1 \\ b_2 \end{bmatrix}.$$

The subscript i on the variables has been omitted for the sake of clarity.

The vector on the left will be referred to as $[P_0]$, the matrix on the right as $[P_j]$, and the vector on the right as $[B]$. Thus the normal equations for the general case of $(m - 1)$ independent variables may be written as

$$[P_0] = [P_j] \cdot [B].$$

Note that the requirements of matrix multiplication are satisfied on the right in that the number of columns in $[P_j]$ is $(m - 1)$ and this equals the number of rows in $[B]$.

We now make use of the concept of the inverse of a matrix, which is discussed in Appendix A. If we premultiply each side of the expression above by $[P_j]^{-1}$, we obtain

$$[P_j]^{-1} \cdot [P_0] = [P_j]^{-1} \cdot [P_j] \cdot [B].$$

Note that the order of multiplication is important.

This expression reduces to

$$[P_j]^{-1} \cdot [P_0] = = [I] \cdot [B] = [B].$$

We have now a solution for the vector of regression coefficients $[B]$ in terms of the known sums of squares and cross products. To complete the solution, however, we still have to obtain $[P_j]^{-1}$.

A number of different methods for obtaining the inverse of a matrix are referenced in the appendix. We shall use the square root method in order to compute the multiple regression equation for the data given in Table 6.8. These data relate to the hypothetical problem of the relationship between population density and distance from the downtown of a city which was considered at the beginning of this chapter. In the present example, however, median family income per tract is treated as the dependent variable, and there are three independent variables: population density, distance, and median number of school years completed. The fact that there is no statement of theory as a basis for this analysis is ignored at this point. We are interested only in the computational steps.

TABLE 6.8.   DATA FOR MULTIPLE REGRESSION PROBLEM

Tract	Median income ($1000)	Pop. density (1000 psm)	Distance (1000 ft)	Median no. school years
1	5.0	27.5	5	9.7
2	4.6	21.0	5	8.7
3	4.0	14.5	5	9.8
4	7.1	20.8	10	12.1
5	5.1	20.0	10	9.3
6	4.7	16.6	10	9.5
7	4.3	11.0	10	8.8
8	7.9	19.1	15	12.5
9	6.2	17.2	15	11.5
10	5.1	10.7	15	16.0
11	4.9	7.8	15	9.0
12	11.1	15.2	20	14.7
13	7.1	11.6	20	12.6
14	4.8	9.8	20	11.1
15	10.8	13.2	25	13.3
16	7.7	10.6	25	12.7
17	6.1	7.9	25	12.6
18	5.3	6.5	25	12.4
19	8.5	9.0	30	12.5
20	7.5	6.7	30	12.4
21	6.3	6.3	30	10.3
22	6.0	3.9	30	11.7
23	7.1	7.9	35	12.3
24	7.0	5.3	35	12.4
25	5.6	1.9	35	9.6
26	13.6	5.3	40	13.2
27	7.5	3.4	40	12.7
28	7.0	3.0	40	12.5
29	12.5	1.7	45	14.5
30	7.6	1.0	45	12.4

The matrix of the sums of squares and cross products in deviation form, $[P_j]$, for the three independent variables in this example is

$$\begin{bmatrix} 1316.91 & -2085.77 & -115.42 \\ -2085.77 & 4396.67 & 320.40 \\ -115.42 & 320.40 & 97.74 \end{bmatrix} \cdot$$

The corresponding vector of cross products for the independent and dependent variables, $[P_o]$, is

$$\begin{bmatrix} -130.18 \\ 493.33 \\ 83.13 \end{bmatrix}$$

The solution for the inverse of the matrix $[P_j]$ by means of the square root method gives

$$[P_j]^{-1} = \begin{bmatrix} 0.003232 & 0.001649 & -0.001589 \\ 0.001649 & 0.001140 & -0.001790 \\ -0.001589 & -0.001790 & 0.014226 \end{bmatrix}.$$

It should be noted that if the original matrix $[P_j]$ is multiplied by this inverse as a check on the computation, then the resulting product matrix only approximates the identity matrix. This is a consequence of rounding error.

Having obtained this inverse, then the multiple regression coefficients are obtained from the matrix equation,

$$[B] = [P_j]^{-1}[P_o].$$

Hence,

$$[B] = \begin{bmatrix} 0.003232 & 0.001649 & -0.001589 \\ 0.001649 & 0.001140 & -0.001970 \\ -0.001589 & -0.001790 & 0.014226 \end{bmatrix} \cdot \begin{bmatrix} -130.18 \\ 493.33 \\ 83.13 \end{bmatrix}$$

$$= \begin{bmatrix} 0.260666 \\ 0.198927 \\ 0.506403 \end{bmatrix}.$$

The constant term a in the multiple regression equation is obtained from the expression

$$a = \bar{X}_0 - b_1\bar{X}_1 - b_2\bar{X}_2 - \ldots - b_{(m-1)}\bar{X}_{(m-1)}.$$

In the present example, we have

$$a = 6.933333 - (0.260666)(10.543332)$$
$$- (0.198927)(23.666666) - (0.506403)(11.759999),$$
$$= -6.478193.$$

The multiple regression equation for the example in Table 6.8 is, therefore,

$$Y = -6.478193 + 0.260666X_1 + 0.198927X_2 + 0.506403X_3,$$

where Y represents the predicted value of the dependent variable X_0.

Standard Partial Regression Coefficients

The solution outlined above for the normal equations gives values for the b's in the multiple regression equation

$$Y = a + b_1 X_1 + b_2 X_2 + \ldots + b_{(m-1)} X_{(m-1)},$$

where Y is again the predicted value of the dependent variable.

It is important to understand the meaning of these b values. They are, in fact, partial regression coefficients in the sense that each one gives the rate of change in the dependent variable for a unit change in the particular independent variable while the remaining independent variables are held statistically constant. The multiple regression equation with three independent variables is written correctly as

$$Y = a + b_{01.23} X_1 + b_{02.13} X_2 + b_{03.12} X_3,$$

where the subscript 0 represents the dependent variable.

The partial regression coefficients are not independent of the particular metrics used for the variables. It is meaningless, therefore, to attempt to compare directly the magnitudes of the b values as in any way indicative of the relative importance of the different variables in the regression equation.

It is easy to standardize the partial regression coefficients, however. The new standard values are represented as the β_j's. They are derived as follows:

$$\beta_j = b_j(s_j/s_0),$$

where s_j is the standard deviation of the independent variable in question and s_0 is the standard deviation of the dependent variable. Since they are dimensionless terms, the β_j's can be compared directly as measures of the relative importance of the different variables in accounting for variation in the dependent variable.

The student reader might compute these standard partial regression coefficients for the example in Table 6.8. The standard deviations in question are

$$s_0 = 2.39, \qquad s_1 = 6.74, \qquad s_2 = 12.31, \qquad s_3 = 1.84.$$

What can be said about the relative importance of the three independent variables in accounting for the variance in the dependent variable?

The normal equation also can be solved in such a way as to yield the β_j's directly rather than the b_j's. If the original variables are in standard form (mean 0, variance 1), the normal equations can be written in terms of the simple correlation coefficients and their solution then gives the standard partial regression coefficients. Cooley and Lohnes (1962) outline this solution. In matrix form it is given as

$$[R_{ij}]^{-1} \cdot [R_{0j}] = [\beta_j],$$

where $[R_{ij}]$ is the matrix of intercorrelations among the $(m-1)$ independent variables and $[R_{0j}]$ is the vector of simple correlations for the dependent variable and each of the independent variables.

In his study of population growth in suburban Chicago, Thomas (1960a) developed a multiple regression model with as many as nine independent variables. The standard partial regression coefficients for these variables are given in Table 6.9, and as shown, the two variables "birth-death differential" and "density of population" contributed far more than the other seven toward explaining the variation in population growth.

TABLE 6.9.   STANDARD PARTIAL REGRESSION COEFFICIENTS FOR
THOMAS'S CHICAGO STUDY

Independent variables	Standard partial regression coefficients
Amount of vacant land	−0.02914
Accessibility of an outlying city	−0.15893
Cost of housing	−0.02494
Birth-death differential	0.40187*
Number of persons engaged in manu- facturing	0.03631
Age of an outlying city	0.10300
Density of the population	−0.73640*
Size of the population	0.35312
Quality of the schools	0.20506

*Significant at the 0.05 level.
Source: Thomas (1960a, p. 165).

Multiple and Partial Correlations

In the multivariate case, the coefficient of multiple correlation $R_{0.123\ldots(m-1)}$ can be derived from the ratio of the sum of squares for regression to the sum of squares for the dependent variable. Hence,

$$R_{0.123\ldots(m-1)} = [\textstyle\sum y^2 / \sum x_0^2]^{1/2},$$

where

$\sum y^2$ = sum of squares for regression
$\qquad = b_1 \sum x_0 x_1 + b_2 \sum x_0 x_2 + \ldots + b_{(m-1)} \sum x_0 x_{(m-1)},$

$\sum x_0^2$ = sum of squares for the total or the sum of squares for the dependent variable. This expression is the same as before in the simple regression analysis.

The coefficient of multiple determination R^2 indicates the proportion of the total variance "explained" by the multiple regression.

We have discussed already the concept of a partial regression coefficient. There remains to be defined the *partial correlation coefficient*. This is the correlation between two variables with either one or more of the other variables held constant. In a problem involving two independent variables, there will be three such partial correlation coefficients. They are defined as follows:

$$r_{01.2} = \frac{r_{01} - r_{02}r_{12}}{[(1 - r_{02}^2)(1 - r_{12}^2)]^{1/2}},$$

$$r_{02.1} = \frac{r_{02} - r_{01}r_{21}}{[(1 - r_{01}^2)(1 - r_{21}^2)]^{1/2}},$$

and

$$r_{21.0} = \frac{r_{21} - r_{20}r_{10}}{[(1 - r_{20}^2)(1 - r_{10}^2)]^{1/2}}.$$

It is quite probable that $r_{01.2}$, for example, will differ markedly from r_{01} and, similarly, $r_{02.1}$ will differ from r_{02}. It may even be the case that the partial correlation between two variables with a third held constant will differ in sign from the simple correlation between the same two variables.

Some of these points regarding partial correlation coefficients are illustrated in Olsson's study of migration which was cited earlier. One of the hypotheses examined in this study was that the length of a migration would be negatively related to the age of a migrant. The rationale behind this hypothesis does not concern us here. The simple correlation coefficient for the two variables concerned was $+.814$, which did not substantiate the hypothesis. When the level of income in the place of outmigration was held constant, however, the partial correlation between migration distance and age of migrant was equal to $-.041$. In other words, for a given income level among the places of outmigration, the hypothesis that migration distance would be negatively related to the age of migrant was substantiated, although not very strongly.

Partial correlations are not interpreted easily, least of all when there are several independent variables involved. It is noteworthy, however, that in some problems of social science research they are emphasized and subjected to very critical examination. This is particularly so in some of the survey research into voting behavior and in many areas of sociological research. Blalock (1964) provides an excellent review of these research designs and discusses the role of partial correlation coefficients in evaluating causal models in nonexperimental situations. He acknowledges that much of his work on causal inference was stimulated by the earlier work of Wold and Jureen (1953) and Simon (1957). Cox (1968) has made a start in applying similar research designs in political geography.

Inferential Questions in Multiple Regression and Correlation

As in the simple two-variable case, there are a number of inferential questions which can be considered in multiple regression and correlation.

The overall significance of the multiple regression can be tested by an analysis of variance. This test is equivalent to testing the null hypothesis that in the population from which the observations have been sampled there is no significant regression of the dependent on the independent variables. The analysis of variance is set up as follows:

Sources	Sum of squares	d.f.	Mean square
Regression	$\sum y^2 = b_1 \sum x_0 x_1 + b_2 \sum x_0 x_2 + \cdots$ $+ b_{m-1} \sum x_0 x_{m-1}$	$m - 1$	$\sum y^2/(m - 1)$
Residual	$\sum u^2 = \sum x_0^2 - \sum y^2$	$N - m$	$\sum u^2(N - m)$
Total	$\sum x_0^2 = \sum (X_{0i} - \bar{X}_0)^2$	$N - 1$	

In this table, m is the total number of variables, including both the dependent and the independents, and N is the number of observations. The F ratio is formed as follows:

$$F = \frac{\sum y^2/(m - 1)}{\sum u^2/(N - m)}, \qquad \text{d.f.} = (m - 1) \text{ and } (N - m).$$

If the computed F exceeds the tabled value for a selected level of confidence, then the null hypothesis is rejected.

Greater emphasis usually is placed on testing the significance of the individual partial regression coefficients. This requires the calculation of the standard errors for these regression coefficients. At this point, we need to refer back to the inverse of the sums of squares and cross products matrix $[P_j]$. The elements of this inverse matrix can be referred to as c_{ij}, where $i = 1, 2, \ldots, (m - 1)$ and $j = 1, 2, \ldots, (m - 1)$. Then the estimated standard error of b_j, written as $\hat{\sigma}_{b_j}$, is $[c_{jj} \sum u^2/(N - m)]^{1/2}$.

The term c_{jj} is the element of the inverse $[P_j]^{-1}$ corresponding to the jth variable in question, and the expression $\sum u^2/(N - m)$ is the mean square for residuals which appeared in the analysis of variance in the table above. The square root of the latter expression is the standard error of estimate for the multiple regression. This is written as $s_{0. 12 \ldots (m-1)}$.

Once $\hat{\sigma}_{b_j}$ has been computed, then the significance of the jth variable in the regression can be tested using a "t" statistic:

$$t = b_j/\hat{\sigma}_{b_j}, \qquad \text{d.f.} = N - m.$$

Similarly, the standard error of the difference between any two b values can be derived as

$$\hat{\sigma}_{b_i - b_j} = [\sum u^2(c_{ii} + c_{jj} - 2c_{ij})/(N - m)]^{1/2},$$

and the significance of the difference between the b values can be tested using

$$t_{b_i - b_j} = \frac{b_i - b_j}{\hat{\sigma}_{b_i - b_j}}, \qquad \text{d.f.} = N - m.$$

The significance of the standard partial regression coefficients can be tested in a similar fashion. The standard errors of the β_j's are

$$\hat{\sigma}_{\beta_j} = [c_{jj}(1 - R^2)/(N - m)]^{1/2},$$

and the "t" statistics are defined as above.

It often is desirable to delete variables which do not appear significant in a regression model. Since they are not contributing to the explanation of variation in the dependent variable, they may indeed appear superfluous. In their place it may be possible to add new variables which had been ignored previously.

Prior to the widespread use of electronic computers in multivariate analysis, the deletion and addition of variables involved considerable hand calculation. The elements of the inverse had to be adjusted and new values obtained for the partial regression coefficients.

McCarty, et al. (1956), for example, in their study of the location of the machinery manufacturing industry in the United States in 1950, found that at the county scale two of the independent variables were not significant. These show up in Table 6.10. The two variables were deleted and the adjusted regression coefficients computed (Table 6.11). As seen from the table, the adjustments were relatively minor and the level of "explained variation" indicated by R^2 virtually was unchanged.

TABLE 6.10.   RESULTS OF MULTIPLE CORRELATION AND REGRES-
SION ANALYSIS, COUNTIES, 1950 (10 VARIABLES)

Variable	β	t	Significant*
Percent of workers in manufacturing	0.1579	6.4449	yes
332	0.2500	6.9252	yes
336	0.1922	5.3241	yes
339	−0.2447	7.0723	yes
344	0.0276	0.8313	no
346	0.2569	4.8562	yes
348	0.2039	4.6766	yes
36	0.3020	7.1226	yes
371	−0.0144	0.4162	no
R = 0.913			yes†
R^2 = 0.8338			

*The least significant value of t is 1.966.
†The least significant value of R with $N = 414$ and $m = 10$ is 0.205.
Source: McCarty, et al. (1956, p. 96).

TABLE 6.11. RESULTS OF MULTIPLE CORRELATION AND REGRESSION ANALYSIS, COUNTIES, 1950 (8 VARIABLES)

Variable	β	t	Significant*
Percent of workers			
in manufacturing	0.1563	6.3796	yes
332	0.2507	6.9446	yes
336	0.1915	5.3047	yes
339	−0.2386	6.8960	yes
346	0.2587	4.8904	yes
348	0.1985	5.5528	yes
36	0.3168	7.4717	yes
R = 0.913			yes†
R² = 0.8335			

*The least significant value of t is 1.966.
†The least significant value of R with $N = 414$ and $m = 8$ is 0.190.
Source: As for Table 6.9.

TABLE 6.12. SIMPLE CORRELATION COEFFICIENTS

	Y	X_1	X_2	X_3	X_4	X_5	X_6	X_7	X_8
Y	1.000	.866	.827	.763	.863	.821	.775	.814	.810
X_1		1.000	.947	.839	.975	.956	.785	.948	.928
X_2			1.000	.687	.959	.964	.762	.932	.900
X_3				1.000	.758	.732	.640	.730	.747
X_4					1.000	.947	.859	.945	.921
X_5						1.000	.711	.936	.904
X_6							1.000	.743	.746
X_7								1.000	.930
X_8									1.000

Y = migration distance.
X_1 = level of income in place of outmigration.
X_2 = level of unemployment in place of outmigration.
X_3 = number of inhabitants in place of outmigration.

X_4 = level of income in place of inmigration.
X_5 = level of unemployment in place of inmigration.
X_6 = number of inhabitants in place of inmigration.
X_7 = age of migrant.
X_8 = income of migrant.

Source: Olsson (1965, p. 18).

Today, a widely used procedure in the computer manipulation of a regression model is the *stepwise* solution. This approach involves adding one independent variable at a time and generating a series of intermediate regression equations. The first independent variable considered is the one which has the

TABLE 6.13 SUMMARY TABLE FOR THE STEPWISE REGRESSION MODEL

Step	Variable	Included variables						Variables not incl.	
		R	Std. err.	R^2	Increase	Regr. coeff.	Std. err.	Variables	Partial corr. coeff.
1	X_1	0.8656*	0.3117	0.7493*	0.7493*	$b_1 = 0.08837^*$	0.00066	X_2 X_3 X_4 X_5 X_6 X_7 X_8	0.04608 0.13765 0.17510 −0.04735 0.30749 −0.04101 0.03907
2	X_6	0.8792*	0.2966	0.7730*	0.0237*	$b_1 = 0.06842^*$ $b_6 = 0.03656^*$	0.00102 0.00147	X_2 X_3 X_4 X_5 X_7 X_8	0.01836 0.16246 −0.04776 0.02309 −0.04044 0.01702
3	X_3	0.8826*	0.2927	0.7790*	0.0060*	$b_1 = 0.05569^*$ $b_3 = 0.02333^*$ $b_6 = 0.03756^*$	0.00142 0.00184 0.00145	X_2 X_4 X_5 X_7 X_8	0.14806 0.06735 0.11242 0.02290 0.04234
4	X_2	0.8853*	0.2895	0.7838*	0.0048*	$b_1 = 0.02181^*$ $b_2 = 0.25075^*$ $b_3 = 0.03923^*$ $b_6 = 0.03630^*$	0.00325 0.02170 0.00228 0.00144	X_4 X_5 X_7 X_8	0.01368 0.03769 −0.00068 0.02688

5	X_5	0.8855*	0.2893	0.7841*	0.0003†	$b_1 = 0.01681*$ $b_2 = 0.20931*$ $b_3 = 0.03990*$ $b_5 = 0.07758†$ $b_6 = 0.03793*$	0.00367 0.02594 0.00229 0.02665 0.00154	X_4 0.00699 X_7 −0.00479 X_8 0.02419
6	X_8	0.8856*	0.2893	0.7843*	0.0001‡	$b_1 = 0.01455*$ $b_2 = 0.20688*$ $b_3 = 0.04011*$ $b_5 = 0.07391*$ $b_6 = 0.03770*$ $b_8 = 0.00704‡$	0.00387 0.02597 0.00229 0.02672 0.00155 0.00377	
7	X_7	0.8856*	0.2893	0.7843*	0.0001‡	$b_1 = 0.01622*$ $b_2 = 0.20876*$ $b_3 = 0.03951*$ $b_5 = 0.07665*$ $b_6 = 0.03763*$ $b_7 = -0.09779†$ $b_8 = 0.00891†$	0.00412 0.02602 0.00234 0.02682 0.00155 0.08267 0.00409	X_4 0.00448 X_7 −0.01533
8	X_4	0.8856*	0.2893	0.7843*	0.0000‡	$b_1 = 0.01458*$ $b_2 = 0.20565*$ $b_3 = 0.03998*$ $b_4 = 0.00269‡$ $b_5 = 0.07432*$ $b_6 = 0.03666*$ $b_7 = -0.10433‡$ $b_8 = 0.00882†$	0.00517 0.02668 0.00251 0.00511 0.02718 0.00241 0.08360 0.00409	X_4 0.00683

*Significant on the 99 percent level.
†Significant on the 95 percent level.
‡Not significant.
Source: Olsson (1965, pp. 10–20).

highest simple correlation with the dependent variable. This initial two-variable regression is completed, and then the partial correlations between the dependent and all the other independent variables are computed. The independent variable among these which has the highest partial correlation, in other words, the one which contributes most to the unexplained variation in the dependent variable remaining after the first regression, is then included at the second step. A new regression equation now involving two independent variables is derived; the partial correlations are computed for the remaining variables with the two held constant; and the selection of the next variable to be included is made on the basis of these values. At each step, the adjusted partial regression coefficients and multiple correlation coefficient also are obtained. The stepwise procedure continues until all the specified independent variables are included.

An example of a stepwise regression, taken from Olsson's migration study, is given in Tables 6.12 and 6.13. The first table reports the simple correlation coefficients for the nine variables. As seen, it is variable 1 that has the highest correlation with the dependent variable Y, and it is this independent variable which is entered first into the stepwise regression (Table 6.13)

6.4   THE MAPPING OF RESIDUALS

The majority of spatial patterns are highly complex and apparently result from the interplay of numerous variables, many of which have still to be identified. Therefore, in most geographic analyses there is considerable unexplained variation and the residual values can be interpreted as reflecting, in part, the effect of other possibly unknown variables. The qualification "in part" should be noted, because even in cases where the correlation between two variables is high, sampling error alone will result in some departure of the observations from the associated regression line. It becomes a question of how to interpret the ϵ values in the model conceptualization. Besides, it is well to remember that the regression line is a plotting of an average relationship; hence the residual values are, in part, mathematical corollaries of this averaging.

The mapping of residual values has been suggested as a valuable tool in subsequent hypothesis formulation. Thomas (1960b) favors the mapping of the standardized residual $(X_{0i} - Y_i)/s_{0.1}$ as the approach to this question. The distribution of these values is symmetrical around zero, which is the residual at the mean value $\bar{X}_0$.

Hopefully, the locational arrangement of residual vaues will suggest other variables which might be important in accounting for the remaining variation in the dependent variable. This demands that we look for some pattern in the distribution of either the underpredicted or overpredicted values and identify other spatial factors which might be associated with this pattern.

This form of analysis is illustrated in a study of transport development in Nigeria and Ghana completed by Taaffe, Morrill, and Gould (1963). The regression model involved two independent variables (total population and land area) and a dependent variable (road mileage) measured for a number of subregional units. After the joint relationship of total population and land area had been considered, the pattern of residuals shown in Figure 6.6 remained. These are nonstandardized residuals $(X_{0i} - Y_i)$. This pattern suggested to the authors "five additional factors: hostile environment; rail competition; intermediate location; income or degree of commercialization; and relationship to the ideal-typical sequence." Each of these factors was considered in turn, and as a consequence some additional findings were made concerning the transport systems of these two countries.

One final point might be made with regard to the analysis of residual values. As noted earlier, the statistical model of regression assumes that these residuals are normally distributed and independent. In the literature of econometrics there are lengthy discussions of this assumption and the forms of analysis which are appropriate, for example, when there is an obvious lack of independence among the error terms. The same problem now is being recognized in geographic research, except that where the economist's observations typically relate to temporal data, the geographer is more concerned with spatial data. This problem is considered in more detail in the final section of this chapter, and only one general comment is entered here. If the residuals do show a strong pattern, then it may be that the explanatory model being considered is not a very powerful one since the variables in it are not accounting fully for the spatial variation in the dependent variable. The difficulty of interpretation in this case, however, may be compounded by the fact that the dependent and/or the independent variables may be characterized individually by built-in spatial correlation effects. These effects, henceforth, will be referred to as *spatial autocorrelations*, and as we shall discuss later, they are an important consideration in any study of spatial relationships.

6.5 APPLICATIONS OF MULTIPLE REGRESSION AND CORRELATION IN GEOGRAPHY

Multiple regression and correlation models have been used widely in geographic research. Some of the earliest applications were by McCarty and his associates (1956) in studying the locational associations of metals industries, Hartman and Hook (1956) in their analysis of substandard housing in U.S. cities, and Garrison (1956b) in his study of the relationships between highway improvement and land values. The more recent literature of urban geography

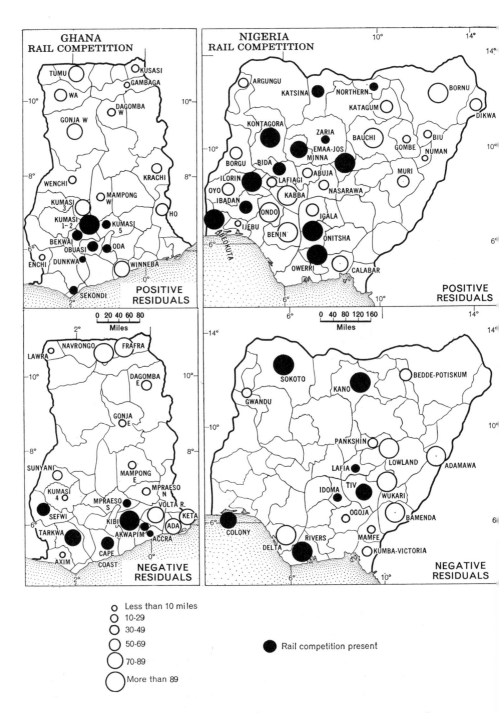

Figure 6.6. Maps of regression residuals. In these maps the possible effect of another variable "rail competition" on the residuals is considered. On all the residuals maps, circles are proportional to the residual deviation

includes a number of studies structured around multiple regression analyses. Thomas's study of population growth in Chicago's suburbs (1960a) has been referred to already. The investigations into urban land values by Knos (1962) and Yeates (1965) employed linear regression models. The same type of model appeared in King's study of the spacing of cities (1961), in Salisbury and Hart's analysis of village population change in the Midwest (1965), in Lakshmanan's study of Baltimore's land-use structure (1964), and in Russwurm's analysis of central business district retail sales (1964).

In broader areas of population geography research, Robinson and his associates (1961) have employed multiple regression analysis in studying rural farm population densities in the Great Plains, and Kariel (1963) has used a similar model in studying county migration patterns. The Swedish geographer Olsson (1965), in his study of migration distances, also employed a multiple regression model with eight independent variables being included. This analysis was reported earlier in the chapter in connection with the stepwise regression procedure.

Additional applications which can be cited include Taaffe's studies of transportation patterns (1958, 1959), Roberts' and Rumage's analysis of voting patterns (1965), Wong's investigation into mean annual floods for the New England drainage basins (1963), and McConnell's study of slope characteristics of stream-dissected glacial materials (1966b).

In all these studies, the emphasis was on establishing the degree and direction of the relationships between a particular spatial distribution considered as the dependent variable and the spatial patterns associated with selected independent variables.

It is a fair commentary on these studies to note that generally the major concerns have been with the level of explained variation as represented in the R^2 and in deciding which variables contributed significantly to the regression. The fact that the latter often is meaningless, given the absence of formal sampling designs in many of the studies, has not been discussed seriously. In addition, there has been no critical discussion of the estimating procedures employed. This is in contrast to the literature of econometrics, where there are lengthy discussions of estimating techniques other than the single-equation procedures which we have reviewed in this chapter. As shown in the final section of this chapter, there are some important technical problems

---

from the regression formula. For example, on the positive residuals maps, the large circles for Tumu, Ghana, and Abeokuta, Nigeria, indicate that they had considerably *more* highway mileage than would be expected from the regression formula, which takes into account population and area. Similarly, on the negative residuals maps the large circles for Kibi, Ghana, and Delta, Nigeria, indicate that they had considerably *less* highway mileage than expected. In effect, then, these units show the distribution of the variation in highway mileage attributable to factors other than population and area. Source: Taaffe, Morrill, and Gould (1963).

present in spatial regression analysis which can undermine the efficiency of this single-equation approach. The future, therefore, may see the use of alternative models in geographic research.

6.6 TREND SURFACE ANALYSIS

A special case of the multiple regression problem which appears relevant to geographic research is *trend surface analysis*. For a map pattern of a variable Z, which is measured at a number of discrete location points, the analysis involves the regressing of this dependent variable on two independent variables U and V, which are orthogonal coordinates for the map area.

This type of analysis has been applied extensively by geologists, and Krumbein and Graybill (1965) provide an excellent review of the approach. They define trend surface analysis "as a procedure by which each map observation is divided into two or more parts: some associated with the 'larger-scale' systematic changes that extend from one map edge to the other and the others associated with 'small-scale' apparently nonsystematic fluctuations that are imposed on the large-scale patterns" (p. 321). In equation form, the model is

$$Z_{ij} = \tau(U_i, V_j) + e_{ij},$$

where Z_{ij} is the observed value of the dependent variable at location (U_i, V_j), $\tau(U_i, V_j)$ is the trend, and e_{ij} is the residual. The problem is to find the parameters of the trend.

Characteristically, this trend is defined as a polynomial involving at least the first-order (linear) and second-order (quadratic) terms and, in some cases, even higher terms. Hence,

$$Z_{ij} = \alpha_{00} + \alpha_{10}U_i + \alpha_{01}V_j + \alpha_{20}U_i^2 + \alpha_{11}U_iV_j + \alpha_{02}V_j^2 + \ldots + e_{ij}.$$

The nature of the solution for the parameters ($\alpha_{00} \ldots \alpha_{02}$) depends upon the relative location of the sample points being considered. In the special case when the points are equally spaced along the U axis and also along the V axis, then the solution is facilitated by the use of standard *orthogonal polynomials* which are tabled. It should be noted that the requirement concerning equal spacing of the sample points does not necessarily imply a square grid, only a regular rectangular one. That is, the interval separating the points on one axis does not have to equal the interval on the other axis.

Krumbein and Graybill (1965, pp. 307–314) outline a simple example concerning the application of orthogonal polynomials, and the reader is referred to this discussion. Hare (1958) also discusses the use of orthogonal polynomials in fitting mathematical surfaces to north polar atmospheric pressure data.

In the case of irregularly spaced sample points, the solution is computationally more involved. The application of the method of least squares results in a set of normal equations, as in the multiple regression problem discussed earlier, except that now these equations involve second- and perhaps even higher-order terms. For example, in fitting a fourth-order or quartic model, the following terms are required:

$$N, \quad \sum U, \quad \sum V, \quad \sum U^2, \quad \sum V^2, \quad \sum UV, \quad \sum U^3, \quad \sum U^2V,$$
$$\sum UV^2, \quad \sum V^3, \quad \sum U^4, \quad \sum U^3V, \quad \sum U^2V^2, \quad \sum UV^3, \quad \sum V^4.$$

It is not surprising, therefore, that this line of analysis has been dependent upon computer-derived solutions, and now there are available programs which will compute and plot trend surfaces up to even the sixth degree. The reader is referred to Whitten (1963) and O'Leary, et al. (1966) for examples of these programs.

Once the parameters for a surface have been obtained, then the significance of the associated trend can be assessed in the same way as the ordinary multiple regression model. The percentage contribution made by each surface to the total sums of squares of the dependent variable can be computed. Krumbein and Graybill (1965, p. 430) have also discussed the setting of confidence intervals around a fitted surface.

The computation and mapping of residual values from a surface affords an opportunity to examine the data for secondary trends. The identification of these secondary trends may be consistent with inferences from a model, or alternatively it may prompt further theorizing as to why they should be present.

Chorley and Haggett (1965) have provided an excellent review of the limited work along these lines which has been attempted for geography. They note that to date most of the trend surface analysis has concerned physical phenomena and continuous surfaces. But they argue that this need not always be the case and that, for example, "population, like light, may be profitably regarded either as a series of discontinuous quanta or as a continuum." The fitting of trend surfaces to social and economic data may provide then a "useful yardstick for describing geographical patterns," and conceivably it may even allow the testing of hypotheses about the spatial patterns of particular variables.

These suggestions have been pursued already in a few studies. Gould (1966c) has suggested some possible applications of trend surface analysis in the examination of spatial perception surfaces. Fairbairn and Robinson (1967a, b) have used the technique to examine the location pattern of urban services among a set of towns and more recently a pattern of residuals from regression. Haggett (1967) has discussed the use of trend surface analysis in "interregional comparison of intra-regional structures."

6.7 SOME RELATED TECHNICAL PROBLEMS IN GEOGRAPHIC RESEARCH

The numerous applications of correlation and regression analysis in geographic research have been referred to in the preceding sections. An overall appraisal of these studies might prompt the general conclusion that the net contributions to geographic knowledge in most cases have been small. This is not to deny that some significant relationships have been discovered, but the validity of broader generalizations derived from many of these studies is weakened seriously by the almost complete disregard of technical problems associated with the regression analysis of spatially distributed data.

Admittedly, in most of the studies attention has been given to the assumptions required for making inferential statements. This concern has focused on the questions of the normality of the parent distributions and the linearity of the relationships, and these assumptions usually have been satisfied by transformations of the study data. There remain, however, other more troublesome assumptions which typically have been ignored. Recently, there have been attempts made to bring these problems into sharper focus, and these discussions will be reviewed in this section.

Modifiable Units

In correlation and regression analyses of data relating to areal units, variation has been observed in the obtained estimates of the regression and correlation coefficients when the sizes of the areal units were altered from one analysis to another.

For example, McCarty, et al. (1956), in their study of the location of the U.S. machinery industry, found that at the county level the simple and multiple correlations were lower than at the state level.

Similarly, Yule and Kendall (1950) noted that the correlation between wheat yields and potato yields for English counties in 1936 varied with the level of data aggregation. When the counties were grouped into larger units, the correlation increased. This prompted the authors to stress that ". . . correlations will . . . measure the relationship between the variates for the specified units chosen for the work. They have no absolute validity independently of those units, but are relative to them" (p. 312).

Robinson (1956) was probably the first geographer to review this problem seriously. His well-known solution involved weighting the calculations for the regression and correlation coefficients by the areas involved. For example, the estimating formula for the regression coefficient b became

$$b = [\sum A_i \sum A_i X_i Y_i - (\sum A_i X_i)(\sum A_i Y_i)]/[\sum A_i \sum A_i X_i^2 - (\sum A_i X_i)^2],$$

where A_i was the area of the ith observation.

Thomas and Anderson (1965) pointed out that this areal weighting provides no general solution to the problem of handling different-sized areal units. Only in very special cases will the weighting remove the size effect in the regression computations.

In a later study, Robinson, et al. (1961) suggested the use of regular hexagons as the units of study in preference to irregularly sized counties. A method of interpolating values for these hexagons is illustrated in Figure 6.7. As an operational procedure, however, this does not represent a very practicable solution to the problem. The interpolation is not only time-consuming and arduous, but there is no guarantee that it will not introduce as much error, perhaps even more, as would result from the use of the original county data.

The discussion of this general problem by Thomas and Anderson (1965) relies heavily on the notions of statistical inference. At the outset, a distinction is made between a *population* as the total set of outcomes or events from which a sample may be drawn and a *universe* which is "a more abstract group" containing "all events as they happened and as they might have happened if everything else had remained the same but the random shocks." We noted this particular distinction in our earlier discussion of sampling. It is a critical one for the Thomas-Anderson solution, since it follows then that "the data for each of. . . several study areas may be treated as a random sample from some hypothetical universe of possible values although each is, in fact, a population." It is feasible, therefore, "to evaluate differences between parameters which describe areal associations within the study area populations by the same methods applied when attempting to evaluate the magnitudes which arise within any random samples, i.e., by employing appropriate tests of statistical significance."

Thomas and Anderson present several hypothetical and actual examples of these inferential tests. The examples all satisfy certain basic assumptions which are demanded by the solution. These assumptions are as follows:

(i)   The study areas have the same total area.

(ii)   The total magnitudes of any two phenomena being considered in a simple regression are equal in the different study areas. That is to say, the sum of W for area 1 equals the sum of W for area 2, and so on.

(iii)   It is possible to define density figures for each areal unit included in a study area.

(iv)   The study areas involve different combinations of the areal units of an initial study area, as in Figure 6.8.

(v)   The paired values of the two variables are randomly assigned throughout a study area.

(vi)   Each study area is a population from some larger universe.

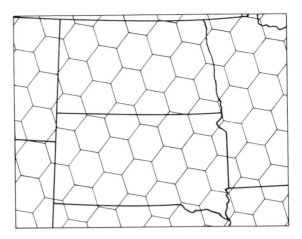

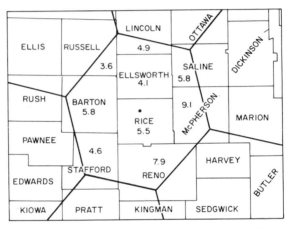

Figure 6.7. Hexagonal grid for data aggregation. A portion of the hexagonal grid in central Kansas showing how county rural farm population density data were converted to unit area values for the hexagons. The numbers are the densities and the data for each hexagon were processed as follows:

County	Density	×	Estimated Portion of hexagon	=	Product
Rice	5.5		0.17		0.93
Ellsworth	4.1		0.17		0.70
Reno	7.9		0.16		1.26
Barton	5.8		0.13		0.75
Stafford	4.6		0.12		0.55
McPherson	9.1		0.12		1.09
Saline	5.8		0.06		0.35
Lincoln	4.9		0.04		0.20
Russell	3.6		0.03		0.11
			1.00		5.94

The resultant density used for the hexagon centered on Rice County is 5.94 persons per square mile. *Source:* Robinson, Lindberg, and Brinkman (1961).

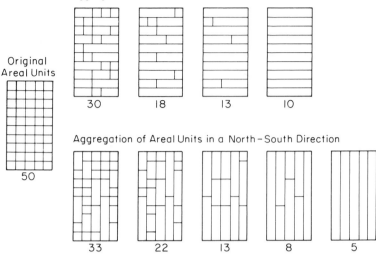

Figure 6.8. Hypothetical data for different aggregations of areal units. The number under each study area refers to the number of areal units within it. *Source:* Thomas and Anderson (1965).

In view of these assumptions, the solution is still not a truly general one. The problem of comparing results obtained from studies based upon completely different sets of data and involving study areas unequal in total size remains unanswered. Thomas and Anderson also point out that there are some related technical questions concerning areal size and the effects of data aggregation which still have to be pursued.

In a recent commentary on the Thomas-Anderson study, Curry (1966b) has questioned the utility of emphasizing the distinction between a population and universe, given the paucity of process knowledge which exists at this time in geography. Curry's point is that "infinite populations can only be defined in relation to processes," and that "the real problems in the study of areal associations are not statistical but rather the dearth of theory on the processes producing the association." Some other related issues discussed by Curry will be referred to shortly.

Spatial Autocorrelation

The linear regression model that we have been discussing up to this point assumes that the error terms in the model

$$X_0 = \alpha + \beta X_1 + \epsilon_i$$

are independent and normally distributed.

In studies where the data relate to contiguous areal units, the assumption of independence is not likely to be satisfied. As Tobler (1966a) has pointed out, "It is not unreasonable to expect that events at one location (i, j) may be related to events at some other geographical location $(i + m, j + n)$. We are thus led to a consideration of serial (or auto-) correlation of each observation with, for example, its neighbor immediately to the east, i.e.,

$$r(Z_{ij}, Z_{i+1, j}),$$

or with the neighbor immediately north

$$r(Z_{ij}, Z_{i, j+1}),$$

or with the neighbor two removed to the east

$$r(Z_{ij}, Z_{i+2, j}),$$

and so on."

The technical consequences of autocorrelation in the use of least squares method for estimating the coefficients in the linear regression model have been discussed by the econometrician Johnston (1963). He notes that there will be three main consequences.

(i)   The sampling variances of the estimates for α and β will not be minimized, and there may also be underestimation of these same sampling variances.

(ii)   The associated inferential tests will no longer be valid.

(iii)   Predictions of the dependent variable will be inefficient in the sense that the sampling variances will be "needlessly large."

Given the econometrician's concern for obtaining efficient estimates, it is not surprising that they have developed an extensive literature on the problem of autocorrelation in time series data.

Unfortunately, geographers have not progressed as far in handling the problem of spatial autocorrelation and only a few scattered references on the topic exist. Some of these are reviewed briefly here.

Geary (1954), in his discussion of contiguity effects which has been referenced in Chapter 5, considered some of the implications for regression analysis. Specifically, he suggested that "if the dependent variables are found to be contiguous, the fact that the remainders after removal of the effect of independent variables are found to lack contiguity constitutes a prima facie case for regarding the independent variables included as completely explaining the dependent variables." The test for contiguity in the spatial arrangement of the residual values which is implied here could be made using Geary's contiguity ratio.

There are, however, some qualifications which need to be entered concerning this question posed by Geary. The first has to do with the other implication in Geary's statement that if there is contiguity in the dependent variable and also in the residuals from regression, then the independent variables considered have low explanatory power. This may be the case, however, only when the spatial trends and patterns of both the dependent variable and the residuals are similar. For it may well be that an independent variable accounts for a reasonably high proportion of the variance in the dependent variable and yet either overpredicts or underpredicts the values in a particular subregion. The residuals, therefore, may still show some contiguity but the spatial trend and pattern of these values may be quite different from that of the dependent variable. A second possibility is that the independent variables also may exhibit strong contiguity effects. This possibility poses very serious problems, and the almost complete disregard for these by geographers has prompted Curry (1966b) to note "that we still really do not know what we are doing in spatial regressions."

Dacey (1965b) also considered the problem of contiguity effects in the spatial arrangement of residuals. The situation was handled as a two-color map problem. Counties which were overpredicted by the regression $(X_0 - Y < 0)$ were considered white; those which were underpredicted or predicted exactly $(X_0 - Y \geqslant 0)$ were black. The test for contiguity was made with respect to the number of different joins present in the map. Two counties were considered to be *contiguous* or *joined* if they had an edge and/or vertex in common. The number of black-black, black-white, and white-white joins could then be counted and these sums represented by Z, Y, and X, respectively.

The expected values for these sums in a random pattern were derived by Dacey. The values are as follows:

$$\mu(Z) = p^2 L$$
$$\mu(Y) = q^2 L$$
$$\mu(X) = 2pqL.$$

In these expressions, p is the probability that a county is black, and $q = (1 - p)$ is the probability that it is white. In a random pattern, $p = q = \frac{1}{2}$. The value of L gives the total number of joins; this equals

$$\frac{1}{2} \sum_{k=1}^{N} L_k,$$

where N is the total number of counties and L_k is the number of counties contiguous to the kth county. The sum of the L_k values is divided by 2 to avoid double counting.

The standard errors of the sums Z, Y, and X are given as follows:

$$\sigma(Z) = [p^2 L + p^3 K - p^4 (L + K)]^{1/2}$$
$$\sigma(Y) = [2pqL + pqK - 4p^2 q^2 (L + K)]^{1/2}$$
$$\sigma(X) = [q^2 L + q^3 K - q^4 (L + K)]^{1/2}.$$

These expressions require K, which is "the total number of counties that have no county in common" (Dacey 1965b, p. 12):

$$K = \sum_{k=1}^{N} L_k(L_k - 1).$$

The test for randomness considers the expressions

$$\frac{Z - \mu(Z)}{\sigma(Z)}, \qquad \frac{Y - \mu(Y)}{\sigma(Y)}, \qquad \frac{X - \mu(X)}{\sigma(X)}$$

as standardized normal variates. The null hypothesis is that there is no significant difference between the observed sums and the expected values. Dacey notes that this test is not sensitive to all forms of autocorrelation. For example, a periodic trend in the residuals may not be picked up on this test which considers only immediately adjacent counties. Also, the abstraction into a two-color map ignores considerable information and it may be desirable to include more than two categories in the analysis.

What seems to be required in studies of spatial association is that we have greater understanding of the autocorrelation effects that Tobler referred to and that were mentioned at the start of this section. This is a frontier area of geographic research, and in the final chapter we shall consider some of the more recent work on this problem. This new work goes beyond the autocorrelation functions into power spectrums and other more advanced topics. At this point, however, we might note that there are few available references to empirical work on spatial autocorrelation functions. Curry (1967) references some work on this problem.

Matern (1960) in his study of forest surveys did present some empirical findings concerning one-dimensional spatial autocorrelation functions (Table 6.13). These values were derived from map analysis of the Stockholm region. The variable of interest, "distribution of land area," was measured on the simple binary scale, an observation being given a value of *1* if on land and *0* if on water. Four different series of autocorrelations were computed.

The first series was based upon the analysis of data obtained by drawing 12 equidistant *E–W* lines on a map of scale 1:250,000 and making 272 observations, equally spaced 1 mm apart on the map, along each of these lines. Then for the first 260 points on each line, counted from the west, the autocorrelations were computed for lags of 1 mm up to 12 mm. In other words, the correlation with a lag of 4 mm, for example, relates to the paired observations of the 260 fixed points and the 260 points located 4 mm to the east of these fixed points. Since there were 12 different lines, the correlation actually was computed from the $12 \times 260 = 3120$ paired observations. In this example, the correlation was equal to 0.491.

The remaining three series of autocorrelations are based on different scale maps. Series 2 was derived as follows:

> From each one of 18 points systematically located in the map, four rays of length 50 mm. were drawn in directions NE, N, NW, and W. On each ray, $Z(x)$ was registered for points at distance 1 mm. (50 meters) apart.

Since a few lines cut the edge of the map, less than 50 points were surveyed in some cases. All pairs on a line with mutual distance k mm. were used for the calculation of a correlation r_k. Thus the correlation coefficients are based on a varying number of point pairs (e.g. 3279 for r_0, and 1945 for r_{20}).

Finally, the third and fourth series "are based on parallel lines running across the map in direction east-west (series 3) and north-south (series 4). Points 10 mm. (500 meters) apart were observed. Also, in these cases all possible pairs of points on a line were utilized when computing the correlation coefficients."

The correlation values for these series have been plotted in Figure 6.9. This graph and Table 6.13 illustrate two of the general features which Matern emphasized with regard to spatial autocorrelation functions. First, the correlation decreases monotonously with increasing distance. Second, the functions can be described "by curves that have negative derivatives at the origin and are downward convex in the vicinity of the origin." Matern suggested that negative exponential curves should apply in this context. Finally, it was noted that the correlations often are "nearly isotropic but may show a certain influence of direction as well as distance."

Analyses similar to that of Matern's have still to appear in the geographic literature. One limiting factor is that many of the patterns in which the geographer is interested are not sufficiently continuous to allow for the generation

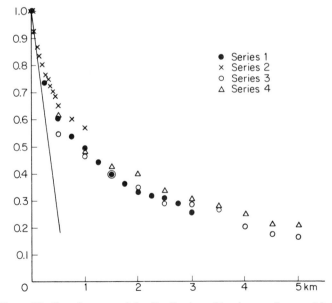

Figure 6.9. Correlograms of the distribution of land area. *Source:* Matern (1960).

TABLE 6.13.   OBSERVED SERIAL CORRELATIONS (r_k) OF THE DISTRI-
BUTION OF LAND AREA IN THE STOCKHOLM REGION

Lag (k)	Series 1 Map on the scale 1:250,000 East–West	Series 2	Series 3	Series 4
		Map on the scale 1:50,000		
		Four direc-tions	East–West	North–South
0	1.000	1.000	1.000	1.000
1	0.736	0.924	0.546	0.605
2	0.601	0.868	0.463	0.485
3	0.537	0.834	0.395	0.423
4	0.491	0.803	0.332	0.393
5	0.442	0.767	0.294	0.331
6	0.397	0.748	0.284	0.303
7	0.361	0.721	0.261	0.277
8	0.328	0.708	0.203	0.245
9	0.316	0.690	0.176	0.207
10	0.308	0.656	0.168	0.205
11	0.285			
12	0.255			
15		0.606		
20		0.572		
Lag 1 corresponds to the interval:				
On map	1 mm	1 mm	10 mm	10 mm
In the field	250 m	50 m	500 m	500 m
No. point pairs for a correlation	3120	1945–3259	2091–2550	2091–2550

Source: Matern (1960, p. 53).

of many data points. In other cases, the data are simply not available for a
series of equally spaced location points that is required in this type of analysis. These considerations will be raised again in the final chapter.

Other Problems in Regression Analysis

The problems of modifiable units and spatial autocorrelation are not
the only ones encountered in regression analysis. For example, there is the
problem of *multicollinearity*, which describes the situation where there are
high correlations among the independent variables. In this case the estimates
of the regression coefficients and partial correlations have large standard

errors and hence little utility in the research. Blalock (1963) has discussed this problem, and Christ (1966) and Johnston (1963) have considered it from a somewhat more technical viewpoint as it relates to econometric research.

In the econometrics literature, multicollinearity appears as a special case of a more general problem, that of *identification*. This involves not only questions of model conceptualization but also technical questions in estimation procedures. Christ (1966, p. 300) notes that "the idea of identifiability refers to the issue of whether the model is just sufficiently restrictive so that when it is confronted with the data . . . , just one hypothesis is consistent with both model and data." Hence, "the concept of identification is usually defined in the literature in terms of whether a structure or a parameter value can be inferred uniquely from the model and . . . data" (Christ 1966, p. 304).

These problems have yet to be posed in the geography literature, even though they appear relevant in the context of many research problems in the discipline. In the central place studies, for example, the problem of considering the relationship between, say, threshold level and number of functional units, given that an increase in population size has occurred, would seem to involve questions of identification.

Another problem relates to the use of *dummy variates* in regression analysis. Both Knos (1962) and Yeates (1965) used such variables in incorporating regional effects into their studies of urban land values. Suits (1957) has noted the problems which may arise in the solution of the normal equations when there are dummy variates in the regression model. Draper and Smith (1966) illustrate the different uses of dummy variates in multiple regression.

An alternative research design to the one used by Knos and Yeates is the *analysis of covariance*. This is essentially a combination of regression analysis and the analysis of variance. Given the regression of X_0 on X_1, a treatment or regional effect is introduced and the observations are divided into groups with reference to this new factor. Then it is of interest to test whether the group X_0 means differ significantly after they have been adjusted for the regression of X_0 on X_1. Also, the contribution of the treatment or regional effect to the level of total explained variation is calculated.

The analysis of covariance proceeds on the basis of these assumptions:

(i)   The groups or regions have equal variances on X_0.

(ii)   The independent variable X_1 is not affected by the regional division. In other words, the group means, $\bar{X}_1$, do not differ significantly. This appears to be a particularly restrictive assumption in the nonexperimental research which is characteristic of much geographic inquiry.

(iii)   The group regression lines are parallel.

These assumptions and the analysis of covariance computations are discussed in Snedecor (1956) and a special issue of *Biometrics* (1957). In the literature of geography, the analysis of covariance has been used by Taaffe

(1958), King (1961), Kariel (1963), and Salisbury and Hart (1965). For the most part, the applications of the technique in these contexts have not been very rewarding. Indeed, it is seldom the case that the assumptions above have even been satisfied in these studies.

SUGGESTED READINGS

Chorley, R. J., and Haggett, P. (1965), "Trend Surface Mapping in Geographical Research." *Publication No. 37*, Institute of British Geographers, pp. 47–67.

Curry, L. (1966b), "A Note on Spatial Association." *The Professional Geographer*, Vol. 18, pp. 97–99.

Curry, L. (1967), "Quantitative Geography." *The Canadian Geographer*, Vol. 11, pp. 265–279.

Fraser, D. A. S. (1958), *Statistics: An Introduction*. New York: John Wiley & Sons, Inc., Chapter 12.

Garrison, W. L. (1956b), *The Benefits of Rural Roads to Rural Property*. Seattle: State Council for Highway Research.

Geary, R. C. (1954), "The Contiguity Ratio and Statistical Mapping." *The Incorporated Statistician*, Vol. 5, pp. 115–145.

Haggett, P. (1964), "Regional and Local Components in the Distribution of Forested Areas in Southeast Brazil: A Multivariate Approach." *Geographical Journal*, Vol. 130, pp. 365–380.

Hoel, P. G. (1960), *Elementary Statistics*, Second edition. New York: John Wiley & Sons, Inc., Chapters 8 and 9.

Johnson, N. L., and Leone, F. C. (1964), *Statistics and Experimental Design*. Vol. I. New York: John Wiley & Sons, Inc., Chapter 12.

Olsson, G. (1965), "Distance and Human Interaction. A Migration Study." *Geografiska Annaler*, Vol. 47, pp. 3–43.

Taaffe, E. J., Morrill, R. L., and Gould, P. R. (1963), "Transport Expansion in Underdeveloped Countries: A Comparative Analysis." *Geographical Review*, Vol. 53, pp. 503–529.

Thomas, E. N. (1960a), "Areal Associations Between Population Growth and Selected Factors in the Chicago Urbanized Area." *Economic Geography*, Vol. 36, pp. 158–170.

Yeates, M. H. (1965), "Some Factors Affecting the Spatial Distribution of Chicago Land Values 1910–1960." *Economic Geography*, Vol. 41, pp. 57–70.

PRINCIPAL COMPONENTS
AND FACTOR ANALYSIS
IN GEOGRAPHIC RESEARCH

Given the complexity of most geographic problems with regard to the number of interrelated variables that have to be considered, it is not surprising that multivariate procedures increasingly are being favored in geographic analysis. Multiple regression analysis has been discussed in the previous chapter, and now we consider two other widely used multivariate statistical methods, namely principal components analysis and factor analysis. Both of these techniques result in the collapsing of a set of intercorrelated variables onto a smaller number of basic dimensions or "composite variables." As such, they largely have replaced in current research the older technique of cluster analysis, which deals with the same general problem.

Although the methods of principal components analysis and factor analysis have much in common and indeed often are referred to as though identical, there are important mathematical and conceptual

7

differences between them. Lawley and Maxwell (1963) emphasize these contrasts in an exposition of the two methods, and their discussion provides the basis for the following summaries of the techniques.

7.1 PRINCIPAL COMPONENTS ANALYSIS AND APPLICATIONS

Principal components analysis essentially involves an orthogonal transformation of a set of variables $(x_1, x_2, \ldots, x_m)$ into a new set $(y_1, y_2, \ldots, y_m)$. The mathematics of an orthogonal transformation and the nature of the solution for the components $(y_1, y_2, \ldots, y_m)$ is taken up shortly. For the moment, however, we note that the transformation results in $(y_1, y_2, \ldots, y_m)$ being uncorrelated one with another, notwithstanding the fact that the original variables $(x_1, x_2, \ldots, x_m)$ may have been quite highly intercorrelated. Also, it is important to note there are as many components derived as there are variables, and the original total variance associated with $(x_1, x_2, \ldots, x_m)$ is preserved exactly in the total variance of the components $(y_1, y_2, \ldots, y_m)$. The solution, moreover, is such that y_1 accounts for the highest proportion of this total variance, y_2 for the second largest share, and so on.

The solution of the principal components analysis was outlined originally by Hotelling (1933). As an introduction to the analysis, however, we rely here upon a simple graphical exposition given by Seal (1964). In Figure 7.1 there is a set of equiprobability ellipses for a bivariate normal distribution. The 50 percent line, for example, is simply a "contour" enclosing 50 percent of the observations in the bivariate normal distribution. In this particular case, the variates X_1 and X_2 are uncorrelated, and the fact that they have

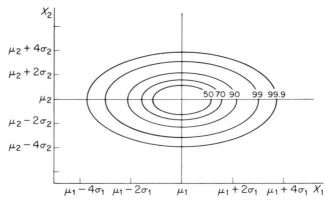

Figure 7.1. Equiprobability ellipses for a bivariate normal distribution. *Source:* Seal (1964).

different variances is reflected in the elliptical rather than circular shape of the equiprobability lines.

The assumption that we are dealing with multivariate normal distributions is not essential for the principal components solution unless certain inferential questions are posed. There is no harm done, however, in retaining the assumption at this time. Of greater interest is the possibility that X_1 and X_2 are correlated. Now the equiprobability ellipses will be rotated according to an angle α, the cosine of which equals the correlation coefficient for X_1 and X_2. Only one of these rotated ellipses is shown in Figure 7.2.

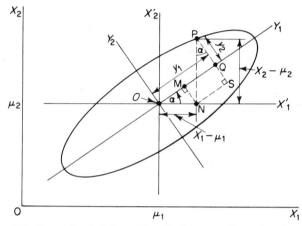

Figure 7.2. An equiprobability ellipse in the case of correlated variates. *Source:* Seal (1964).

Consider an observation or point P with coordinates $(X_1 - \mu_1)$ and $(X_2 - \mu_2)$ in Figure 7.2. We wish to transform this point with respect to two new axes, Y_1 and Y_2, which are orthogonal and uncorrelated. That is, $r_{12} = 0 = \cos \alpha$, where $\alpha = 90°$.

From the diagram in Figure 7.2, we note that

$$y_1 = OM + MQ.$$

Now

$$OM = (X_1 - \mu_1) \cos \alpha$$

and

$$MQ = NS = (X_2 - \mu_2) \sin \alpha.$$

Therefore,

$$y_1 = (X_1 - \mu_1) \cos \alpha + (X_2 - \mu_2) \sin \alpha.$$

Similarly,

$$y_2 = PS - SQ.$$

But

$$PS = (X_2 - \mu_2) \cos \alpha,$$

and

$$SQ = MN = (X_1 - \mu_1) \sin \alpha.$$

Therefore,

$$y_2 = (X_2 - \mu_2) \cos \alpha - (X_1 - \mu_1) \sin \alpha$$
$$= -(X_1 - \mu_1) \sin \alpha + (X_2 - \mu_2) \cos \alpha.$$

Combining the expressions above in matrix form, we have

$$\begin{bmatrix} y_1 \\ y_2 \end{bmatrix} = \begin{bmatrix} \cos \alpha & \sin \alpha \\ -\sin \alpha & \cos \alpha \end{bmatrix} \cdot \begin{bmatrix} X_1 - \mu_1 \\ X_2 - \mu_2 \end{bmatrix}$$
$$= \begin{bmatrix} \cos \alpha & \sin \alpha \\ -\sin \alpha & \cos \alpha \end{bmatrix} \cdot \begin{bmatrix} x_1 \\ x_2 \end{bmatrix},$$

with the last equation simply relating to the X values in deviation form.

From these equations it is possible to derive expressions for the x values as linear transforms of the y's. That is,

$$\begin{bmatrix} x_1 \\ x_2 \end{bmatrix} = \begin{bmatrix} \cos \alpha & \sin \alpha \\ -\sin \alpha & \cos \alpha \end{bmatrix}^{-1} \cdot \begin{bmatrix} y_1 \\ y_2 \end{bmatrix}.$$

This derivation is consistent with our earlier matrix manipulation of a set of equations. We now have to consider what the nature of the inverse matrix is:

$$\begin{bmatrix} \cos \alpha & \sin \alpha \\ -\sin \alpha & \cos \alpha \end{bmatrix}^{-1}.$$

If we multiply the matrix

$$\begin{bmatrix} \cos \alpha & \sin \alpha \\ -\sin \alpha & \cos \alpha \end{bmatrix}$$

by its own transpose, we obtain

$$\begin{bmatrix} \cos \alpha & \sin \alpha \\ -\sin \alpha & \cos \alpha \end{bmatrix} \cdot \begin{bmatrix} \cos \alpha & -\sin \alpha \\ \sin \alpha & \cos \alpha \end{bmatrix}$$
$$= \begin{bmatrix} \cos^2 \alpha + \sin^2 \alpha & -\cos \alpha \sin \alpha + \sin \alpha \cos \alpha \\ -\sin \alpha \cos \alpha + \cos \alpha \sin \alpha & \sin^2 \alpha + \cos^2 \alpha \end{bmatrix}$$
$$= \begin{bmatrix} 1 & 0 \\ 0 & 1 \end{bmatrix},$$

since

$$\sin^2 \alpha + \cos^2 \alpha = 1,$$

and the other elements reduce to zero.

Recall that we noted earlier that the product of a matrix and its own inverse equals the identity matrix. In the present case the transposed matrix apparently also satisfies this condition, and we conclude that here the transpose and the inverse are the same matrix. This is a general property of any *orthogonal* matrix $[A]$, that is, $[A]^T = [A]^{-1}$. Also, it is easily shown with regard to the orthogonal matrix

$$[A] = \begin{bmatrix} \cos \alpha & \sin \alpha \\ -\sin \alpha & \cos \alpha \end{bmatrix}$$

that

$$[A] \cdot [A]^T = [A]^T \cdot [A] = [I]$$

and

$$[A] = [A^T]^{-1}.$$

In rewriting our earlier equations concerning the x's and y's, we shall use $[U]$ as the designation for the matrix

$$\begin{bmatrix} \cos \alpha & -\sin \alpha \\ \sin \alpha & \cos \alpha \end{bmatrix}.$$

Hence $[x] = [U] \cdot [y]$ and $[y] = [U]^T \cdot [x]$.

It can be shown that the matrix containing the variance and covariance terms for the y's is given by the expression

$$[U]^T \cdot \sum \cdot [U],$$

where $\sum$ is the variance-covariance matrix of the original variables. Lawley and Maxwell (1963, p. 46) and Anderson (1958, pp. 272–279) discuss the derivation of this variance-covariance matrix for the y's. Since the y's are assumed to be uncorrelated, the off-diagonal elements in $[U]^T \cdot \sum \cdot [U]$, in other words the covariance terms, will be zeros. On the principal diagonal there will be the variances of the y's, and these are represented usually as λ's.

Continuing with the two-variable example, then, we have

$$[U]^T \cdot \sum \cdot [U] = \begin{bmatrix} \lambda_1 & 0 \\ 0 & \lambda_2 \end{bmatrix} = [\Lambda].$$

Postmultiplication by $[U]^T$ gives

$$[U]^T \cdot \sum \cdot [U] \cdot [U]^T = [\Lambda] \cdot [U]^T$$

and, hence,

$$[U]^T \cdot \sum = [\Lambda] \cdot [U]^T$$

since $[U]$ is an orthogonal matrix and $[U] \cdot [U]^T = [I]$.

If instead of sines and cosines the elements of $[U]$ now are given in general form as

$$\begin{bmatrix} u_{11} & u_{12} \\ u_{21} & u_{22} \end{bmatrix},$$

then we can write the equation $[U]^T \cdot \sum = [\Lambda] \cdot [U]^T$ in the following form:

$$\begin{bmatrix} u_{11} & u_{21} \\ u_{12} & u_{22} \end{bmatrix} \cdot \begin{bmatrix} \sigma_{11} & \sigma_{12} \\ \sigma_{21} & \sigma_{22} \end{bmatrix} = \begin{bmatrix} \lambda_1 & 0 \\ 0 & \lambda_2 \end{bmatrix} \cdot \begin{bmatrix} u_{11} & u_{21} \\ u_{12} & u_{22} \end{bmatrix},$$

or

$$\begin{bmatrix} u_{11}\sigma_{11} + u_{21}\sigma_{21} & u_{11}\sigma_{12} + u_{21}\sigma_{22} \\ u_{12}\sigma_{11} + u_{22}\sigma_{21} & u_{12}\sigma_{12} + u_{22}\sigma_{22} \end{bmatrix} = \begin{bmatrix} \lambda_1 u_{11} & \lambda_1 u_{21} \\ \lambda_2 u_{12} & \lambda_2 u_{22} \end{bmatrix}.$$

Subtracting the right-hand side from the left gives

$$\begin{bmatrix} u_{11}(\sigma_{11} - \lambda_1) + u_{21}\sigma_{21} & u_{11}\sigma_{12} + u_{21}(\sigma_{22} - \lambda_1) \\ u_{12}(\sigma_{11} - \lambda_2) + u_{22}\sigma_{21} & u_{12}\sigma_{12} + u_{22}(\sigma_{22} - \lambda_2) \end{bmatrix} = \begin{bmatrix} 0 & 0 \\ 0 & 0 \end{bmatrix}.$$

The first row of this new expression contains two equations, both equal to zero and involving the two unknowns u_{11} and u_{21}. In order for these equations to have a nonzero solution, then it must be the case that

$$\left\| \begin{bmatrix} \sigma_{11} - \lambda_1 & \sigma_{21} \\ \sigma_{12} & \sigma_{22} - \lambda_1 \end{bmatrix} \right\| = 0,$$

and similarly for u_{12} and u_{22}

$$\left\| \begin{bmatrix} \sigma_{11} - \lambda_2 & \sigma_{21} \\ \sigma_{12} & \sigma_{22} - \lambda_2 \end{bmatrix} \right\|$$

must equal zero. If we ignore the subscripts on the λ's and combine these two equations, we have

$$\left\| \begin{bmatrix} \sigma_{11} - \lambda & \sigma_{21} \\ \sigma_{12} & \sigma_{22} - \lambda \end{bmatrix} \right\| = 0 = |\sum - \lambda I|.$$

The expression on the right-hand side is a "characteristic equation." The reader should consult the section on matrix algebra in Appendix A for an introduction to this topic. Gould (1966b) has provided an excellent review of the related mathematical concepts and their application in geographic research.

   In the simple two-variable case we are considering, the characteristic equation can be solved as follows:
The determinant

$$\begin{vmatrix} \sigma_{11} - \lambda & \sigma_{21} \\ \sigma_{12} & \sigma_{22} - \lambda \end{vmatrix} = (\sigma_{11} - \lambda)(\sigma_{22} - \lambda) - \sigma_{12}\sigma_{21}$$

$$= \lambda^2 - \lambda(\sigma_{22} + \sigma_{11}) + \sigma_{11}\sigma_{22} - \sigma_{12}\sigma_{21}.$$

We have here an expression in λ, which has to equal zero. That is,

$$\lambda^2 - \lambda(\sigma_{22} + \sigma_{11}) + \sigma_{11}\sigma_{22} - \sigma_{12}\sigma_{21} = 0.$$

The solution of this quadratic equation yields two values of λ, which are known as the "eigenvalues," the "characteristic roots," or the "latent roots."

If we recall that in the solution of a simple quadratic equation, $x^2 - bx + c = 0$, the coefficient b equals the sum of the roots, then for the equation in λ given above we note that $b = (\sigma_{22} + \sigma_{11}) = \lambda_1 + \lambda_2$. Hence, the original total variance is preserved by the orthogonal transformation.

Once the values of λ are determined, the equations given earlier in matrix form allow us to solve for the u values. The larger of the two λ values is taken as λ_1, the smaller as λ_2. Then from the equation $[y] = [U]^T \cdot [x]$, the y values can be obtained.

As they appear at this stage, the y's are uncorrelated but nonstandardized. In order to obtain standardized components $(y'_1, y'_2, \ldots, y'_m)$, that is, components with variances equal to one, we have to scale each y by the square root of the corresponding eigenvalue. Hence,

$$y'_r = \lambda_r^{-1/2} \cdot y_r,$$

or in matrix form,

$$[y'] = [\Lambda]^{-1/2} \cdot [y] = [\Lambda]^{-1/2} \cdot [U]^T \cdot [x].$$

It is possible to express the original x's in terms of these standardized components. Premultiplying the equation above by $[U] \cdot [\Lambda]^{1/2}$ gives

$$[U] \cdot [\Lambda]^{1/2} \cdot [y'] = [U] \cdot [\Lambda]^{1/2} \cdot [\Lambda]^{-1/2} \cdot [U]^T \cdot [x],$$

and thus

$$[U] \cdot [\Lambda]^{1/2} \cdot [y'] = [x].$$

This follows since $[\Lambda]$ is a diagonal matrix and the product of $[\Lambda]^{1/2} \cdot [\Lambda]^{-1/2}$ is the identity matrix. Similarly, the product $[U] \cdot [U]^T = [I]$ since $[U]$ is orthogonal.

Now if we adopt the convention of Lawley and Maxwell in setting $[W] = [U] \cdot [\Lambda]^{1/2}$, then we have

$$[W] \cdot [W]^T = [U] \cdot [\Lambda]^{1/2} \cdot [U]^T \cdot [\Lambda]^{1/2}$$
$$= [U] \cdot [\Lambda] \cdot [U]^T.$$

Recall that earlier we took as given the equation

$$[U]^T \cdot \textstyle\sum \cdot [U] = [\Lambda].$$

Premultiplying each side of this by $[U^T]^{-1}$ and at the same time postmultiplying by $[U]^T$ gives

$$[U^T]^{-1} \cdot [U]^T \cdot \textstyle\sum \cdot [U] \cdot [U]^T = [U^T]^{-1} \cdot [\Lambda] \cdot [U]^T.$$

Since $[U]$ is orthogonal, this reduces to

$$\textstyle\sum = [U^T]^{-1} \cdot [\Lambda] \cdot [U]^T$$
$$= [U] \cdot [\Lambda] \cdot [U]^T$$
$$= [W] \cdot [W]^T.$$

Further, we note that

$$[W]^T \cdot [W] = [U]^T \cdot [\Lambda]^{1/2} \cdot [U] \cdot [\Lambda]^{1/2}$$
$$= [U]^T \cdot [\Lambda] \cdot [U]$$
$$= [\Lambda].$$

The problem in principal components analysis, then, often is stated as one of finding a matrix $[W]$ such that

$$[W] \cdot [W]^T = \Sigma,$$

and

$$[W]^T \cdot [W] = [\Lambda].$$

If the original variables $(x_1, x_2, \ldots, x_p)$ are standardized, that is, with zero means and unit variances, then the variance-covariance matrix Σ becomes the matrix of simple correlations $[R]$. Thus we require that

$$[W] \cdot [W]^T = [R].$$

In this case, since the variables are standardized with variances all equal to *one*, the total variance equals p, the number of variates. The eigenvalues, in turn, will sum to p.

A generalization of the principal components solution is now outlined in terms of the standardized variables and correlation coefficients. Given the expression $[W]^T \cdot [W] = [\Lambda]$, we can complete the matrix multiplication on the left and obtain a series of equations of the following form:

$$w_{11}^2 + w_{21}^2 + \cdots + w_{p1}^2 = \lambda_1 = V_1,$$
$$w_{12}^2 + w_{22}^2 + \cdots + w_{p2}^2 = \lambda_2 = V_2,$$
$$w_{1p}^2 + w_{2p}^2 + \cdots + w_{pp}^2 = \lambda_p = V_p.$$

Since the first component is to have the highest variance among the set of components, the expression V_1 has to be maximized. But in choosing the coefficients $(w_{11}, w_{21}, \ldots, w_{p1})$ which will maximize V_1, we have also to satisfy the constraints given in the expression $[W] \cdot [W]^T = [R]$; namely that

$$w_{11}^2 + w_{12}^2 + \cdots + w_{1p}^2 = r_{11},$$
$$w_{11}u_{21} + w_{12}w_{22} + \cdots + w_{1p}w_{2p} = r_{12},$$

and so on.

We have here a problem involving the maximization of V_1, a function of several variables which in turn are connected by a set of relationships. The mathematical procedure as outlined in Harman (1960, pp. 155–157) involves the use of Lagrangian multipliers to obtain a set of p equations of the form

$$
\begin{bmatrix}
1-\lambda & r_2 & \cdots & r_{1p} \\
r_{21} & 1-\lambda & \cdots & r_{2p} \\
\vdots & \vdots & \vdots & \vdots \\
r_{p1} & & \cdots & 1-\lambda
\end{bmatrix}
\begin{bmatrix}
w_{11} \\
w_{21} \\
\vdots \\
w_{p1}
\end{bmatrix} = 0.
$$

For this to have a nonzero solution, we have the same condition as before; namely,

$$|R - \lambda I| = 0.$$

Corresponding to the first root or eigenvalue of this equation is a column vector or eigenvector $(a_{11}, a_{21}, \ldots, a_{p1})$, which when scaled by the factor $[\lambda_1/(a_{11}^2 + a_{21}^2 + \cdots + a_{p1}^2)]^{1/2}$ yields the coefficients $(w_{11}, w_{21}, \ldots, w_{p1})$. The residual correlation matrix $[R']$ can then be computed as

$$[R'] = [R] - [w_{j1}] \cdot [w_{j1}]^T \qquad \text{for } j = 1, 2, \ldots, p,$$

and the solution could proceed with the finding of the largest eigenvalue of this residual matrix, and so on.

It turns out that these subsequent solutions are unnecessary since each of the successive eigenvalues of the original $[R]$ matrix proves to be the largest eigenvalue of the corresponding residual matrix. Anderson (1958, p. 276) gives a proof of the related theorem.

The availability of high-speed computers has facilitated the solution for the eigenvalues and eigenvectors. Many of the programmed solutions, for example, that one given in Cooley and Lohnes (1962) yield eigenvectors which already are normalized in the sense that the squared coefficients sum to unity, and the w_{ij} coefficients are obtained simply by scaling these eigenvectors by $\lambda_j^{1/2}$.

The solution outlined above relates to the model

$$[Z] = [W] \cdot [y'],$$

where $[Z]$ is the vector of standardized original variables, $[y']$ is the vector of standardized components, and $[W]$ is the orthogonal matrix of coefficients. It follows that

$$[y'] = [W]^T \cdot [Z].$$

This equation allows the computation of *component scores* for the original N observations. For any observation i, there is a column vector of component scores

$$[y_k'] = [W]^T \cdot [Z_j] \qquad \text{for } j, k = 1, 2, \ldots, p,$$

where $[Z_j]$ is the column vector of values on the standardized original variables for the ith observation.

It also is easy to show that the coefficients in the matrix [W] are, in fact, the correlations between the original variables and the components. Harman (1960, pp. 16–17) discusses this feature of the analysis. These coefficients are often referred to as the "loadings" of the variables on the components.

It is important to note that the component loadings are not invariant under a change of the measurement scales for the variables. The mathematical consequences of such changes are discussed in Anderson (1958, p. 279). We note here only the point made in most technical discussions to the effect that "principal components analysis is most useful when the variates x_i are all measured in the same units." The uses of ranked data, or standardized variates, are partial answers to this problem.

The interpretation of the principal components usually is important in a research problem. This interpretation is done mainly with reference to the loadings, the philosophy being that variables having high correlations or loadings on a component will serve to identify that component. Oftentimes in geographic research, the interpretation is aided by a mapping of the component scores. A note of caution, however, is appropriate concerning the component scores. On any component some variables will have low loadings and consequently will be ignored in the process of giving an "interpretation" to the component. On these particular variables, some observations almost certainly will have high values in the original data matrix and their scores on the component will be weighted accordingly by these "unimportant" variables. Again, the use of ranked data helps avoid this issue. Otherwise, the interpretation of the component scores may have to be qualified by an examination of the original data matrix.

The question of how many components should be interpreted is a vexing one. Bartlett has derived tests for the hypothesis that the $(p - k)$ eigenvalues remaining after the k largest ones have been extracted are equal. Lawley and Maxwell note, however, that these tests are not exact in the case when standardized variates are being analyzed. A convenient rule of thumb seems to be to evaluate all components with eigenvalues equal to or greater than one or, alternately to evaluate each one which accounts for "a sufficiently high proportion," say, at least 5 percent, of the total variance.

For an example of the computational results associated with principal components analysis, we refer to the data in Table 7.1. These are the 1940 agricultural data for Ohio used by Weaver (1954) as part of his study of crop-combination regions in the Midwest. The data matrix has 88 observations (counties) and 7 variables.

The variable means are as follows:

$$\bar{X}_1 = 31.78, \quad \bar{X}_2 = 0.14, \quad \bar{X}_3 = 20.16, \quad \bar{X}_4 = 10.39,$$
$$\bar{X}_5 = 0.16, \quad \bar{X}_6 = 5.97, \quad \bar{X}_7 = 25.37.$$

The standard deviations are

$$S_1 = 8.67, \quad S_2 = 0.11, \quad S_3 = 5.68, \quad S_4 = 6.46,$$
$$S_5 = 0.20, \quad S_6 = 6.73, \quad S_7 = 13.78.$$

TABLE 7.1. AGRICULTURAL LAND-USE DATA FOR OHIO COUNTIES, 1949

County	Percent of total harvested cropland (including noncommercial farms)						
	Corn	Mixed small grains	Wheat	Oats	Barley	Soybeans	Hay
Adams	42.41	0.21	22.47	1.07	0.37	0.62	27.80
Allen	34.43	0.13	23.76	18.35	0.11	12.18	15.31
Ashland	27.05	0.02	24.81	17.75	0.07	2.21	24.15
Ashtabula	22.88	0.24	13.52	15.67	0.02	1.30	38.89
Athens	26.61	0.18	8.89	3.42	0.05	0.71	53.91
Auglaze	34.66	0.06	22.02	18.24	0.10	9.80	11.97
Belmont	18.60	0.11	10.87	9.16	0.31	0.20	50.82
Brown	47.04	0.07	22.81	0.97	0.07	2.33	16.30
Butler	44.97	0.29	24.76	3.98	1.28	1.57	23.00
Carroll	19.07	0.23	17.73	16.35	0.12	0.10	43.48
Champaign	43.61	0.15	23.46	12.74	0.09	5.48	20.65
Clark	37.67	0.15	23.22	6.95	0.28	5.43	16.32
Clermont	42.89	0.10	16.58	0.91	0.05	5.00	21.58
Clinton	48.45	0.24	29.50	3.10	0.25	2.72	9.85
Columbiana	20.63	0.14	19.22	17.75	0.26	0.07	31.56
Coshocton	31.79	0.09	21.96	6.05	0.13	0.47	34.78
Crawford	29.61	0.04	18.40	14.72	0.10	12.05	15.67
Cuyahoga	19.25	0.20	12.52	11.78	0.15	4.31	24.19
Darke	42.40	0.10	21.09	12.32	0.05	5.30	16.09
Defiance	25.37	0.22	22.05	18.82	0.04	16.40	12.50
Delaware	33.52	0.13	17.60	11.33	0.16	11.82	22.69
Erie	28.18	0.12	22.11	13.27	0.25	11.87	10.67
Fairfield	40.48	0.01	30.98	2.90	0.10	1.93	23.36
Fayette	41.63	0.14	26.65	5.75	0.27	7.59	11.79
Franklin	35.76	0.39	22.64	6.82	0.29	13.61	19.47
Fulton	38.61	0.14	20.97	20.26	0.07	5.05	14.36
Gallia	31.38	0.83	13.07	2.03	0.60	0.71	44.07
Geauga	23.04	0.21	12.68	17.44	0.11	0.41	37.80
Greene	49.23	0.23	29.30	4.63	0.31	1.97	15.78
Guernsey	22.30	0.18	14.14	3.70	0.08	1.20	54.73
Hamilton	39.25	0.00	13.86	1.60	1.08	3.68	25.06
Hancock	36.13	0.12	24.64	16.56	0.13	13.91	16.46
Hardin	33.07	0.15	16.63	14.61	0.13	13.18	14.18
Harrison	19.30	0.06	13.00	11.56	0.38	0.50	53.62
Henry	31.75	0.09	20.09	18.41	0.08	3.48	10.42
Highland	6.65	0.11	31.57	1.59	0.05	1.46	16.10
Hocking	40.06	0.25	21.46	2.05	0.15	0.48	4.79
Holmes	28.59	0.17	26.38	15.31	0.61	0.16	30.66
Huron	27.70	0.17	21.77	16.33	0.04	13.57	14.88
Jackson	30.07	0.18	14.34	2.88	0.04	0.92	42.29
Jefferson	17.23	0.22	15.61	13.17	0.07	0.20	49.45
Knox	33.75	0.05	26.12	8.71	0.04	1.89	26.20
Lake	16.21	0.07	9.77	10.53	0.18	4.67	22.77
Lawrence	29.85	0.02	4.42	1.26	0.17	0.27	44.72
Licking	33.63	0.14	22.43	6.25	0.05	4.78	24.73

TABLE 7.1 (*Cont.*)

County	Percent of total harvested cropland (including noncommercial farms)						
	Corn	Mixed small grains	Wheat	Oats	Barley	Soybeans	Hay
Logan	34.33	0.22	17.93	15.57	0.17	9.20	20.96
Lorraine	23.15	0.02	16.99	13.53	0.03	14.74	21.84
Lucas	27.97	0.20	18.61	12.58	0.04	15.70	9.50
Madison	38.15	0.28	22.57	8.33	0.27	15.01	13.90
Mahoning	23.22	0.13	20.13	18.12	0.26	0.13	28.67
Marion	34.47	0.13	18.29	14.44	0.06	15.90	15.67
Medina	26.73	0.17	20.86	18.79	0.14	2.84	26.78
Meigs	28.20	0.28	14.08	3.06	0.18	0.67	46.71
Mercer	33.29	0.00	20.48	19.13	0.17	10.39	14.94
Miami	41.76	0.09	26.59	7.69	0.03	5.98	15.08
Monroe	21.35	0.12	9.75	5.54	0.00	0.31	59.79
Montgomery	42.66	0.07	25.88	5.72	0.19	1.50	19.02
Morgan	27.02	0.20	14.94	2.65	0.16	0.40	52.53
Morrow	30.92	0.09	19.26	13.11	0.05	11.03	22.70
Muskingum	31.74	0.11	19.58	3.68	0.25	0.55	40.57
Noble	21.17	0.15	9.24	2.19	0.01	0.24	65.88
Ottawa	21.36	0.15	17.71	9.45	0.11	27.44	13.94
Paulding	23.16	0.31	20.04	16.35	0.00	26.49	7.53
Perry	32.63	0.04	27.96	1.57	0.08	0.67	33.42
Pickaway	41.82	0.07	32.24	3.32	0.02	4.75	14.53
Pyke	48.17	0.01	18.32	3.72	0.00	1.31	26.20
Portage	26.67	0.11	19.13	18.67	0.03	0.69	27.33
Preble	47.26	0.16	26.77	7.18	0.07	1.01	16.49
Putnam	30.97	0.13	24.16	15.28	0.13	14.10	12.61
Richland	27.68	0.03	23.50	17.92	0.06	5.16	21.74
Ross	45.92	0.13	30.24	1.46	0.05	1.55	18.13
Sandusky	30.37	0.09	20.96	14.24	0.16	15.20	14.29
Scioto	44.45	0.23	13.14	1.37	0.23	2.23	25.56
Seneca	31.31	0.04	25.51	14.38	0.12	10.76	15.43
Shelby	36.53	0.05	22.33	15.42	0.03	9.88	14.96
Stark	24.78	0.20	22.53	19.98	0.20	0.18	26.92
Summit	24.37	0.16	19.08	16.49	0.13	0.79	26.58
Trumbull	23.91	0.21	16.53	16.42	0.12	0.74	35.48
Tuscarawas	26.54	0.08	22.49	10.72	0.07	0.56	35.62
Union	32.92	0.23	17.53	12.48	0.09	15.34	19.76
Van Wert	30.00	0.08	18.55	17.13	0.04	22.01	9.02
Vinton	33.68	0.03	11.96	3.01	0.02	0.40	44.96
Warren	43.23	0.09	24.97	3.20	0.24	4.68	18.72
Washington	25.08	0.08	13.43	1.96	0.66	1.06	50.27
Wayne	29.61	0.16	27.10	16.31	0.06	1.20	23.25
Williams	31.71	0.18	21.76	22.37	0.07	4.03	18.10
Wood	38.69	0.14	26.30	21.30	0.04	22.07	13.22
Wyandot	31.36	0.15	22.31	9.91	0.13	19.12	14.14

Source: U.S. Census of Agriculture, 1949.

In Table 7.2 the lower half of the symmetric matrix of simple correlations between the variables is given. The computation of this correlation matrix is the starting point in most computer programs for the principal components analysis. The computer-derived solution to the equation

$$|R - \lambda I| = 0$$

in our example yields the following eigenvalues:

$$\lambda_1 = 2.32, \quad \lambda_2 = 1.69, \quad \lambda_3 = 1.13, \quad \lambda_4 = 0.69,$$
$$\lambda_5 = 0.64, \quad \lambda_6 = 0.37, \quad \lambda_7 = 0.16.$$

The sum of these values is 7.0, which is the trace or the sum of the values on the principal diagonal of the matrix [R] in Table 7.2. This value of 7.0

TABLE 7.2.   MATRIX OF INTERCORRELATIONS

Variable	X_1	X_2	X_3	X_4	X_5	X_6	X_7
X_1	1.00						
X_2	−0.03	1.00					
X_3	0.46	−0.12	1.00				
X_4	−0.33	−0.08	0.08	1.00			
X_5	0.15	0.25	−0.04	−0.26	1.00		
X_6	0.01	0.00	0.13	0.38	−0.19	1.00	
X_7	−0.45	0.11	−0.64	−0.32	0.10	−0.60	1.00

is also, of course, the sum of the variances of the original variables in standardized form. Remember, in other words, that the correlation matrix [R] is the variance-covariance matrix for these standardized variables.

If each eigenvalue is expressed as a percentage of this total variance, then we obtain the results given in Table 7.3.

TABLE 7.3.   PERCENTAGE OF VARIANCE ACCOUNTED FOR BY EACH EIGENVALUE

Eigenvalue	Percent of total variance	Accumulated percent
1	33.17	33.17
2	24.12	57.29
3	16.15	73.44
4	9.93	83.37
5	9.15	92.52
6	5.25	97.77
7	2.23	100.00

The loadings of the seven variables on the seven components are given in Table 7.4. This is the matrix [W].

Recall that the principal components analysis solution is such that

$$[W]\cdot[W]^T = [R] \qquad \text{and} \qquad [W]^T\cdot[W] = [\Lambda].$$

Therefore, from the first row of Table 7.4 we have, for example,

$$[0.45 \quad 0.76 \quad -0.10 \quad -0.18 \quad -0.11 \quad -0.39 \quad 0.11]\cdot\begin{bmatrix} 0.45 \\ 0.76 \\ -0.10 \\ -0.18 \\ -0.11 \\ -0.39 \\ 0.11 \end{bmatrix} = 0.9988$$

$$\approx r_{11} = 1.00.$$

Similarly,

$$[0.45 \quad 0.76 \quad -0.10 \quad -0.18 \quad -0.11 \quad -0.39 \quad 0.11]\cdot\begin{bmatrix} -0.23 \\ 0.16 \\ 0.83 \\ -0.37 \\ 0.32 \\ -0.02 \\ -0.01 \end{bmatrix} = -0.0268$$

$$= r_{12}.$$

Finally,

$$[0.45 \quad -0.23 \quad 0.73 \quad 0.41 \quad -0.26 \quad 0.65 \quad -0.93]\cdot\begin{bmatrix} 0.45 \\ -0.23 \\ 0.73 \\ 0.41 \\ -0.26 \\ 0.65 \\ -0.93 \end{bmatrix} = 2.3114$$

$$\approx \lambda_1 = 2.32.$$

The discrepancies again are due to rounding errors.

The question of interpreting the components in this example can not be pursued to any great length. It is to be noted that in the correlation matrix in Table 7.2 there are very few high intercorrelations, and we might suspect that some of the variables are almost independent of one another in any case. Hence, strong groupings of variables on one or more of the components is unlikely. Furthermore, the example relates to data in per-

TABLE 7.4. MATRIX OF COMPONENT LOADINGS

Variable	Component	I	II	III	IV	V	VI	VII
1	(corn)	0.45	0.76	−0.10	−0.18	−0.11	−0.39	0.11
2	(small grains)	−0.23	0.16	0.83	−0.37	0.32	−0.02	−0.01
3	(wheat)	0.73	0.38	−0.13	0.11	0.44	0.30	0.14
4	(oats)	0.41	−0.70	0.19	0.34	0.28	−0.32	0.08
5	(barley)	−0.26	0.55	0.47	0.61	−0.18	0.04	0.01
6	(soybeans)	0.65	−0.37	0.38	−0.15	−0.47	0.16	0.16
7	(hay)	−0.93	−0.09	−0.14	−0.04	0.05	0.01	0.30

centage form and the correlations may reflect, in part, the biases associated with any such closed number system. Krumbein (1962) provides an excellent discussion of the latter point. The example, however, was presented here since it does serve to illustrate the use of principal components analysis in obtaining orthogonal variates which can then provide a sound basis for a classification or regionalization. The results obtained here will be utilized further to derive a set of agricultural regions for Ohio in the following chapter.

Notwithstanding the points made above, we can attempt some interpretation of the loadings in Table 7.4. The following seven "crop dimensions" seem to emerge:

(i) Component I—wheat, soybeans, corn, and oats—a general crop dimension.

(ii) Component II—corn and barley.

(iii) Component III—mixed grains and barley.

(iv) Component IV—barley and oats.

(v) Component V—wheat and mixed grains.

(vi) Component VI—wheat.

(vii) Component VII—hay.

On each of these seven components, scores for the 88 counties were obtained. These scores are not reproduced here, but in Figures 7.3 and 7.4 those for the first two components have been mapped. It should be noted that the scores are standardized, having a mean of *zero* and variance *one*.

It is obvious how the interpretation of a component might be facilitated by such a mapping. In the present example, the first component separates the counties in the west and northwest from those in the east and southeast portions of the state. The differentiation is based on the greater relative importance of crop farming in comparison to hay farming in the west and northwest. The strong negative loading on this first component for hay farming results in low scores for counties having high percentages of their har-

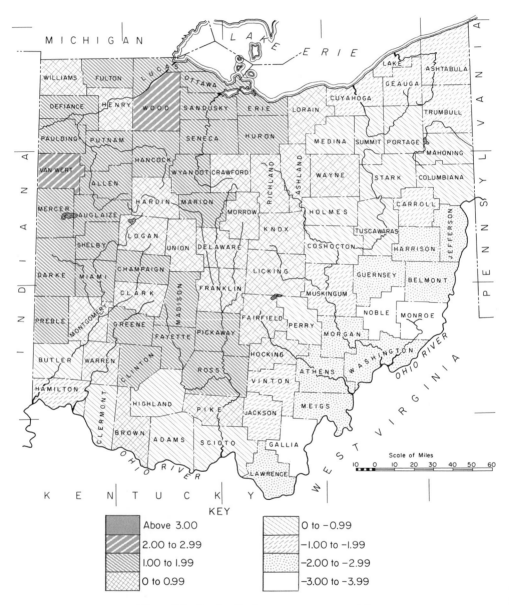

Figure 7.3. Standardized scores on component I.

Figure 7.4. Standardized scores on component II.

vested cropland given over to this form of land use. Such is the case with the counties in the east and southeast of Ohio.

Similarly, on component II the strong negative loading for oats interacts with the high positive loadings for corn and barley to produce the pattern of scores shown in Figure 7.4. Most of the counties in the northern part of the state have reasonably high percentages of their cropland given over to corn, but they also have some of the highest percentages of cropland in oats. As a consequence, they have the lower scores on this component.

There are very few formal applications of principal components analysis in the geographic literature. Oftentimes, the mathematical solution of the principal components analysis is followed, but the overall model conceptualization is more like a factor analysis. Such is the case, for example, with the studies of Wong (1963), Hodge (1965), and Ahmad (1965). In these studies a principal components-type solution was obtained but the "components" were then rotated to new positions. As we shall discuss in the next section, *rotation* is a question that is more properly considered along with factor analysis; therefore, we shall not discuss these particular studies any further at this point.

Moser and Scott (1961) used a principal components analysis in studying the interrelationships among 57 socioeconomic variables measured over some 157 British towns. Their findings were that the first four components together accounted for 60 percent of the original total variance. These components were identified as measuring social class, population change 1931–51, population change 1951–58, and overcrowding.

Although the Moser and Scott study is not part of the geographic literature, it has influenced certain lines of research in geography. The study of Ahmad (1965) is a case in point, and a recent study of Canadian urban dimensions by King (1966) is in the same vein. In the latter study, a principal components analysis was made first of selected 1951 data for 106 cities and, second, of comparable 1961 data for the same cities. The results showed for each year that over 80 percent of the original total variance associated with around 50 variables could be accounted for by 11 or 12 components. For both the 1951 and 1961 analyses, however, meaningful interpretations could be given only to the first 6 components. The 6 components for 1951 were interpreted as indexing the following urban dimensions:

    (i)   The youthfulness of the female populations in the cities, particularly the nonmanufacturing cities.

    (ii)   The service role of many comparatively isolated cities in provinces other than Quebec.

    (iii)   Frontier location and a closer economic orientation to primary activities.

    (iv)   Manufacturing in smaller cities outside the major metropolitan complexes.

(v) Occupational and housing characteristics of suburban communities.

(vi) Metals manufacturing and depressed housing.

These components did not appear very stable over time and were reflected only very weakly in the 1961 results. The components for the 1961 analysis were identified as indexing

(i) Urban manufacturing, especially in textiles.

(ii) Quebec demographic structure.

(iii) Metropolitan socioeconomic structure.

(iv) The residential role of certain communities.

(v) The service function of many older and comparatively isolated cities.

(vi) Depressed urban structure.

Although the urban dimensions or components were not stable over the decade encompassed by this study, some regional groupings of the cities with respect to their scores on the components did appear remarkably stable. This tendency toward order and persistence in the regional differentiation of cities was contrasted with other types of order observed in studies of urban systems.

Casetti (1964a), in an example of the use of another multivariate technique—discriminant analysis—which we still have to discuss, did a principal components analysis of 25 demographic, sociological and economic variables measured for 19 Italian regions. The primary aim in this case was to obtain an orthogonalization of the original data. For the first four components, however, which together accounted for 79 percent of the variance, meaningful interpretations were possible. The components apparently measured

(i) The standard of living.

(ii) The position of the regions in a continuum, with a predominance of tourist activities and mountainous terrain at one extreme (regions with high positive scores) and the predominance of industrial agriculture (high negative scores) at the other.

(iii) The position of the regions in a continuum, at the extremes of which are a predominance of intensive peasant agriculture (high positive scores) and again a predominance of industrial agriculture (high negative scores).

(iv) The contrasts between regions at a high level of economic activities (high positive scores) and those characterized more by poorer economic activities and outward migration.

The use of principal components analysis as an orthogonal transformation of variables which are to be used in a regionalization problem is illustrated further in Stone (1960) and in a study of Poland by Brown and

Trott (1968). The study by Thompson, et al. (1962) of "economic health" in New York also might be regarded as a principal components analysis, although the authors chose to refer to it as a principal axes factor analysis. The distinction implied here is largely one of model conceptualization, and in the case of the Thompson study the lack of a priori hypotheses and attempts to rotate the axes suggest that it is better considered as a principal components rather than a factor analysis. The first three components in this study accounted for three-quarters of the total variance and they were interpreted as "a general economic health scale," "a rural-urban scale," and "a dimension of economic-demographic growth." The Bell and Stevenson (1964) investigation of economic health in Ontario was modeled after this New York study.

As an orthogonalization procedure, principal components analysis often is used in conjunction with regression analysis, when the assumptions of independence and additivity among the independent variables are not satisfied in the latter model. Massey (1965) has discussed at some length the role of principal components analysis in this context. Wong (1963) and Hodge (1965) adopted this strategy, although, as we noted earlier, both chose to rotate their components to new positions prior to using them in the regression.

7.2 THE FACTOR ANALYSIS MODEL

In factor analysis we are again concerned with finding a matrix $[V]$ such that $[V] \cdot [V]^T = [R]$. The matrix $[V]$ contains the factor loadings or correlations between the variables and factors.

However, whereas in the principal components analysis the basic model was of the form

$$z_j = \sum_{k=1}^{p} w_{jk} y'_k \qquad \text{for } j, k = 1, 2, \ldots, p,$$

and the original correlation matrix was reproduced exactly by the product $[W] \cdot [W]^T$, now the factor analysis model is

$$z_j = \sum_{r=1}^{m} v_{jr} f_r + e_j \qquad \text{for } j = 1, 2, \ldots, p; m \leqslant p,$$

and the correlation matrix $[R]$ is only partially reproduced by the product $[V] \cdot [V]^T$. Wherein lies the discrepancy? It is with respect to the terms lying along the principal diagonal of the correlation matrix. In the principal components solution these values were all ones, but now in the reproduced correlation matrix given by $[V] \cdot [V]^T$ most, if not all, of these values will be less then one.

The discrepancy stems from the specification of the e_j term in the basic factor model given above. In this sense the problem is essentially one of model conceptualization and, as Cattell (1965) has noted, there is exposed here "the Achilles' heel of factor analysis."

Factor analysis, as Lawley and Maxwell (1963) emphasize, usually implies some hypothesis as to the number of common factors underlying the set of variables in the research problem. Armstrong (1967) has stressed the dangers inherent in the use of factor analysis when no such hypothesis or theory exists. The variance of every variable then is seen as involving some common variance accounted for by these factors, that is, the *communality* h_j^2, plus some variance associated with a factor e_j specific to the variable in question. Hence,

$$\sigma_j^2 = h_j^2 + \sigma_{e_j}^2.$$

Psychometricians usually insist on a distinction between two contributing influences in the $\sigma_{e_j}^2$ term. One is an error-of-measurement term for which $\sigma_{e_j}^2$ is generally reserved, and the other is the specific variance $\sigma_{u_j}^2$ "that rightly belongs to common factors yet unrepresented." Since in most social science research, and certainly in geography, we have little knowledge or experience as a basis for separating these two influences, here we shall use the term $\sigma_{e_j}^2$ to represent the joint effect.

It is the communalities which appear along the principal diagonal of the correlation matrix in a factor analysis. Usually these estimated values are inserted in the matrix [R] at the start of the analysis, and the mathematical solution can be pursued in the same way as the principal components solution, with a set of eigenvalues and eigenvectors being extracted. This is known as the *principal axes solution*. Other solutions which may or may not be orthogonal have been developed and are described in a number of texts, including Harman (1960), Lawley and Maxwell (1963), and Horst (1965).

Obviously the estimates of the communalities used and the number of factors to be extracted are not independent questions since they both relate to the issue of when the residual correlation matrix becomes zero. In discussing this point, Cattell (1965) suggests that "the better solution is to start by fixing the number of factors" and then estimating the communalities. He notes at the same time, however, a statistical solution which by use of maximum likelihood procedures "goes to and fro between communalities and number of factors until it hits on the combination which yields a smallest residual."

These questions have not been explored in the geography literature. Indeed, to the extent that factor analysis has been used in geography, the principal axes solution has been preferred with estimates of the communalities given by the square of the multiple correlation coefficient between each variable and all the other variables in the problem set. Such is the case, for example, in the studies by Berry (1965) and Berry and Ray (1966), to be discussed shortly.

What can be said about these estimates of the communalities? Obviously, if we use *ones* as our estimates, we are assuming that all the variance of each variable is common to the factors underlying the observed set. If we think of principal components analysis as a form of factor analysis and not simply as an orthogonalization of a set of variates, then this is what we are doing. Cattell (1965, pp. 198–199) dismisses this "closed model" as being "intrinsically unlikely to fit nature," arguing that "in an *R* of, say, 50 or even 100 variables, the correlation of any variable with itself due to the common factors represented among those variables is likely to be decidedly short of unity." The problem is whether this discrepancy is attributable, as Cattell insists, to other "common factors which will only appear when an extremely large number of variables has been coupled with it in a correlation matrix" (p. 199), or to stochastic influences which will never be accounted for in terms of *common* factors. For most geographic situations, I favor the latter possibility.

But if *ones* are not used as the communalities, what estimates are available? As noted in an earlier paragraph, the square of the multiple correlation of each variable with all the others has been favored in geographic applications of the technique. Cattell insists that this is but a lower limit for any communality and that the true value lies between this lower limit and unity. A review of many of the communality estimates which have been proposed is given in Harman (1960, Chapter 5).

Two other important questions arise in factor analysis: first, the matter of factor *rotations* and, second, the need for an estimation procedure to obtain factor scores.

From a mathematical viewpoint, the rotation of factors can be viewed as a consequence of the indeterminacy of the solution. Referring back to Figure 7.1, for example, we can see that a number of different positions could have been chosen for the set of orthogonal axes Y_1 and Y_2. The same is true of the factor analysis solution and the problem is one of deciding which positions are to be preferred. Certain suggestions as to a "simple structure" wherein each factor affects only a limited number of variables and, in turn, each variable is correlated with only a few of the factors generally have been followed. These are only qualitative statements, however, and to facilitate computer handling of the problem, some analytical procedures have been developed. The most commonly used of these is the Varimax routine, which by a series of orthogonal transformations of pairs of factors seeks to simplify the columns of the factor loadings matrix $[V]$. Harman (1960, pp. 301–308) discusses the method in detail. It is this rotational procedure which has been favored in geographic research.

Rotation of the factors naturally changes the loadings of the variables on these variables. The communalities remain the same, however.

One of the unfortunate consequences of the wider availability of computer programs for factor analysis is that rotation of the factors often is performed quite uncritically. Admittedly, given the present wide gaps in understanding which are characteristic of many geographic problem situations, it would be extremely difficult to specify in advance the number of factors to be extracted and the nature of the variables which should have zero or nonzero loadings on the different factors. These are the requirements for some more recently outlined rotational schemes given by Lawley and Maxwell (1963). It should not be impossible to give greater attention to some alternative rotational schemes. Cattell (1965, p. 209) notes:

> Experience in locating naturally-existing hyperplanes, as soon as sufficient scientific data became available, showed, however, what would have been theoretically expected, namely, that in most scientific fields factors are oblique.

The possibility that correlated or oblique factors might be more meaningful in the context of geographic research warrants consideration.

The second important question which we must consider with respect to factor analysis is the computation of factor scores. This computation is not as straightforward as the problem of obtaining principal component scores, since in the factor analysis only a part of the total variance of the original variables is accounted for by the common factors. In other words, a simple reciprocal relationship between the x and y scores which was present in principal components analysis no longer holds true.

Lawley and Maxwell (1963) outline two methods for the estimation of factor scores. Their discussion of the "regression" method is summarized here.

It is desired to find the linear regression of f_r, a common factor, on $(x_1, x_2, \ldots, x_p)$. As in any such regression problem this will require the covariances for f_r and the x variables and the variance-covariance terms for the x's. The latter are given in $\sum$, and as estimates of the former we use the loadings given in $[V]^T$. Hence the estimates for f_r, written as $\hat{f}_r$, are

$$\hat{f}_r = [v_{ir}, v_{2r}, \ldots, v_{pr}] \cdot \begin{bmatrix} \sigma_{11} & \cdots & \sigma_{1p} \\ \cdot & & \cdot \\ \cdot & & \cdot \\ \cdot & & \cdot \\ \sigma_{p1} & \cdots & \sigma_{pp} \end{bmatrix}^{-1} \begin{bmatrix} x_1 \\ \cdot \\ \cdot \\ \cdot \\ x_p \end{bmatrix},$$

or, in matrix form,

$$[\hat{f}] = [V]^T \cdot \sum{}^{-1} \cdot [x].$$

The x's here are assumed to be in standardized form. Lawley and Maxwell (1963, pp. 13–14, 89) show that the expression on the right-hand side of the equation above is equal to

$$(I + J)^{-1} \cdot [V]^T \cdot [R'] \cdot [x],$$

where the residual matrix $[R'] = \sum - [V] \cdot [V]^T$ and $J = [V]^T \cdot [R']^{-1} \cdot [V]$. Thus:

$$[\hat{f}] = [I + J]^{-1} \cdot [V]^T \cdot [R'] \cdot [x].$$

7.3   FACTOR ANALYSIS OF GEOGRAPHIC MATRICES INVOLVING DIFFERENT MEASUREMENT SCALES

In this section, applications of factor analysis in geographic research are reviewed with respect to the type of measurement scales appearing in the data matrix.

First, there have been studies by Berry and Barnum (1962), Garrison and Marble (1964,) and Henshall and King (1966) which involved factor analyses of binary-choice matrices, that is, *nominal* data with only *ones* and *zeros*. The correlation matrix in this case may be a matrix of phi coefficients, although (as we shall note shortly) it is not always necessary to compute this matrix as an intermediate step. Horst (1965, Chapter 22) discusses three different procedures for analyzing a binary data matrix by way of a factor analysis.

Ordinal data have been used in factor analysis studies by Berry (1960a); Schnore (1961); Thompson, et al. (1962); Bell and Stevenson (1964); and Gould (1966c). The use of this type of data was suggested earlier in this chapter as a way around such problems as the presence of different measurement scales for the variables in a principal components analysis and the possibility of undue weighting of the component scores by "extreme values." On the other hand, a reliance upon ordinal data appears to preclude any inferential tests based on the results, given that these tests are derived generally upon the assumption of a multinormal population. These tests are seldom employed in a rigorous manner, however. Also, there are undoubtedly some important mathematical consequences of the use of ordinal data in this context, but these have yet to be considered seriously. Certainly there would seem to be problems in interpreting the component scores derived from such an analysis, since these are now values derived from a linear combination of ranks. An observation, for example, may rank low on a particular variable and have a rank value of, say, 100 or 150. If the variable in question has a high positive loading on a particular factor, then for the observation mentioned above the score on this factor may be quite high. But to the extent that this high value reflects the weighting of the variable above, then it is spurious.

Factor analysis of *interval* and *ratio* scale data has been more characteristic of geographic research. In this context, the urban and regional studies by Berry (1963, 1965, 1966) and his associates have been outstanding. For the most part, these studies have used factor analysis as an important step in the classification and regionalization of observations, the strategy being that the complexity of interrelationships in the original data matrix is reduced by the factor analysis and a fewer number of orthogonal variates can be used in place of the original set. The study by Berry (1963) of the commercial structure of Chicago illustrates this approach. First, the data for some 63 business centers on 10 variables were transformed into common logarithms to ensure linear relationships among the variables. The correlation matrix containing estimates of the communalities given by the multiple coefficients of determination (R^2) was then factor analyzed by the principal axes method. The resulting factors were rotated to a normal Varimax position, and the scores of the observations on these rotated factors then were computed. These scores provided the basis for the subsequent classification. We shall take up this matter of classification in the next chapter. For the moment we note that in this particular study by Berry only two factors stood out: the first "focusing upon size, functions, and sales of centers and the second upon an extra source of variation based on the area, population, and incomes of centers' trade areas" (p. 230).

Basically this same research design has been used in a number of regionalization studies. In one by Berry and Ray (1966), data for the 120 counties of Ontario and Quebec on 88 cultural, demographic, housing, agricultural, urban, manufacturing, and income variables were factor analyzed. Three basic dimensions appeared after rotation to the normal Varimax position. These were

   (i)   Differences between French and English Canada.
   (ii) Variations on an urban-rural scale.
   (iii) Contrasts between the northern shield and southern lowlands.

In a study of rural poverty in Ontario Berry (1965) analyzed data on some 47 variables for 555 census townships and municipalities. Again, the final results suggested three significant basic patterns. The first indexed the differences in wealth and well-being among farmers, with the poorer areas in eastern Ontario showing up at one end of the scale and the St. Lawrence and Ottawa valleys and southern Ontario appearing at the other. The second factor indexed the differences in wealth among the rural nonfarm population, with relative prosperity being marked around the major cities and recreational areas and poverty being more conspicuous in the inter-urban peripheries. The final factor related to "social disadvantage and cultural deprivation," and this served to differentiate the French communities in eastern Ontario from the remaining areas.

Another set of regional studies in which factor analysis is used extensively is that associated with the work of Megee (1963, 1964, 1965a, b). The approach is not significantly different from that in Berry's studies, although the issue of testing hypotheses by way of factor analysis is made somewhat more explicit. Also, in Megee (1965b) there is an attempt to compare the results of a factor analysis for two different points in time.

Mabogunje (1965) used factor analysis in examining some of the demographic impacts of colonial economic development on the Nigerian urban system. The analysis was made of data for 329 towns and 11 variables. The rotated results allowed the identification of three important factors which together accounted for nearly 80 percent of the variance. These factors were interpreted as relating to the proportion of youths or children, the level of older-aged population, and the proportion of females. The patterns of town scores on these factors were then examined in the light of certain hypotheses concerning regional economic development.

Other geographic studies which include factor analyses of interval or ratio data are those of Ahmad (1965) and Wong (1963) which were referred to earlier, Cox (1968) and Carey (1966). Krumbein and Graybill (1965), Kendall (1957), and Seal (1964) reference some applications of factor analysis in the fields of geology, economics, and biology, respectively. Specific studies which are worthy of mention include Imbrie (1963, 1964) and Goodall (1954). In the field of urban and regional planning there have been studies by Hodge (1965, 1966), Fisher (1966), and Gittus (1964) which further illustrate the application of factor analysis to a variety of problems. The work of Russett (1968) and Rummel (1963) in political science also is relevant to some areas of geographic research.

Finally in this section on the factor analysis of sets of different types of measurements, reference is made to a study by Henshall (1966) in which nominal and ratio scale measurements were combined. Of the 32 variables used in this study of Barbadian agriculture, 15 were measured on a binary scale, while the other 17 were measured on a ratio scale. The matrix of correlations in this case included ordinary Pearsonian coefficients for the pairs of ratio variables, biserial correlations for the pairs of ratio and binary variables, and phi coefficients for the pairs of binary variables. This poses no constraint on the mathematical solution, but it could make interpretation difficult. Henshall was able to give meaningful interpretations, however, to the first 12 rotated factors, which apparently accounted for most of the total variation.

7.4 ALTERNATIVE MODES OF FACTOR ANALYSIS APPLIED TO GEOGRAPHIC MATRICES

The discussion up to this point has been concerned with a data matrix in which there are N rows corresponding to the number of observations and p columns representing the variables. This matrix is not a square matrix since it is usually the case that $N > p$. The analysis in this instance proceeds from the correlation matrix $[R]$ to obtain the matrix of loadings, which in the principal components analysis is $[W]$ and in the factor analysis is $[V]$. The matrix $[W]$ is a $p \times p$ matrix; whereas $[V]$ is a $p \times r$ matrix, where r the number of factors is equal to or less than p. The matrix of observation scores on the new dimensions has, of course, N rows in either case but p columns in principal components analysis and r columns in factor analysis.

As an alternative to the design above, consider first a square matrix $N \times N$ in which the cells contain measures of the level of interaction on some feature between the paired observations. These data could be migration flows, transportation movements, or social interactions. The observations, in turn, might be cities, countries, regions, and so on. Such a matrix might be termed a *transactions matrix*, with the rows representing the observations as *origins* and the columns as *destinations*.

A conventional factor analysis of this matrix would proceed as follows: First, the correlation matrix would contain measures of the similarity between the destination vectors with regard to the origins of their "imports." The factor analysis would identify composites or types among these destinations, and the loadings would show which particular destinations were more typical of each type or grouping. Finally, the factor scores matrix would have the N origins as the rows and these destination types or groups as the columns. Certain origins would have high scores on some destination groupings and low scores on others.

Now consider another approach to factoring this transactions matrix. Instead of collapsing the matrix columnwise as suggested above, we could collapse the number of rows. This approach is a Q-mode factor analysis, and it can be applied to any form of data matrix. The more conventional approach, emphasizing the correlations between the column vectors, is known as the R-mode factor analysis.

What does the Q-mode analysis yield in connection with the transactions matrix? Obviously the rows will be collapsed into a set of origin types or exporting groups. The loadings will give the identity of the more important origins associated with each group, and the scores in turn will show which destinations are more important for each group.

Berry (1966) has used both of these approaches in analyzing the commodity flow patterns among 36 trading blocks in India. McConnell (1967) has used the same design in studying patterns of trade in the Middle East.

It is worth noting also that Q-mode analysis was used by Henshall and King (1966) in their study of Barbadian agriculture. In this case it allowed the identification of certain farm types. Imbrie (1963) has discussed the use of Q-mode analysis in geological research.

Saunders (1950) has described a *direct factor* method whereby both the conventional R-mode and the Q-mode results are obtained together. As with many of the factor analysis solutions, there is an iterative procedure involved. This technique is described by Berry and Barnum (1962, p. 45), with reference to the analysis of a binary or "incidence" matrix, as follows: Given the data matrix $[X]$, then "an initial row vector d_1 is selected and the successive operations $Xd_1^T = c_1$; $c_1^T X = d_2$; $Xd_2^T = c_2$; $c_2^T X = d_3 \ldots$ performed until $XX^T c_i = c_{i+1}$ converges at a desired level of accuracy. At this time also $d_i X^T X = d_{i+1}$. The resulting vectors d and c are row and column vectors of $X^T X$ and XX^T, respectively." In the case of the incidence matrix $[X]$, it is easily shown that the sums of the elements along the principal diagonals of $X^T X$ and XX^T are equal and that the total variance which is to be explained is the total number of incidences in $[X]$, specifically $\sum_i \sum_j x_{ij}^2$. Furthermore, "the value c_{ir}^2 indicates the number of incidences of observation i accounted for by component r, and similarly d_{rj}^2 is the number of incidences of variable j attributable to component r." Then,

$$\sum_{i=1}^N c_{ir}^2 = \sum_{j=1}^p d_{rj}^2 = \lambda_r;$$

that is,

$$c_r^T c_r = d_r d_r^T = \lambda_r.$$

Now if I_i and I_j are the row and column sums of incidences for observation i and variable j, respectively, then $\sum_r c_{ir}^2/I_i$ is the communality for observation i, $\sum_r d_{rj}^2/I_j$ is the communality for variable j, $[c_{ir}^2/I_i]^{1/2}$ is the loading of i on r, and $[d_{rj}^2/I_j]$ is the loading of j on r.

Berry and Barnum (1962) use this method in considering further the question of whether or not hierarchies of central places and functions are identifiable. The problem was posed as follows: The incidence matrix $[X]$ with business centers in the rows and central functions in the columns was prepared. "Application of direct factor analysis to X should then extract the effect of the continuum on the first pair of components. If there is a hierarchy, subsequent components should identify the appropriate classes of centers and functions as interaction effects in a series of group factors. Measures of relative significance will be provided by the eigenvalues, and the factor loadings will provide measures of association for individual centers and functions." The results of the Berry and Barnum study are too detailed to report here, but they did bear out the hypothesis stated above.

Berry (1966) has posed one other possibility about factor analysis of geographic matrices. Assume that the rows of the data matrix now contain pairs of places or regions. Each pair is referred to as a *dyad*, and for N places there are $(N^2 - N)$ such dyads. The columns represent different forms of interaction such as wheat shipments and steel movements. An R-mode factor analysis of this matrix will identify "the basic types of spatial interaction and their patterns of areal functional organization" (Berry 1966, p. 191). The factor score matrix will have $(N^2 - N)$ rows and b columns which define the behavior of the system. This matrix is referred to by Berry as the *behavior matrix*. In the last chapter we shall refer to some more advanced forms of analysis in which this matrix features.

SUGGESTED READINGS

Armstrong, J. Scott (1967), "Derivation of Theory by Means of Factor Analysis or Tom Swift and His Electric Factor Analysis Machine." *The American Statistician*, Vol. 21, pp. 17–21.

Berry, B. J. L. (1960a), "An Inductive Approach to the Regionalization of Economic Development." In *Essays on Geography and Economic Development*, Research Paper 62, Department of Geography, University of Chicago.

Berry, B. J. L. (1965), "Identification of Declining Regions: An Empirical Study of the Dimensions of Rural Poverty." In R. S. Thoman and W. D. Wood (eds.), *Areas of Economic Stress in Canada*. Kingston: Queen's University.

Cattell, R. B. (1965), "Factor Analysis: An Introduction to Essentials I and II." *Biometrics*, Vol. 21, pp. 190–215, 405–435.

Cooley, W. W., and Lohnes, P. R. (1962), *Multivariate Procedures for the Behavioral Sciences*. New York: John Wiley & Sons, Inc., Chapter 8.

Gould, P. R. (1966b), "On the Geographic Interpretation of Eigenvalues." Mimeographed paper, Department of Geography, Pennsylvania State University. Forthcoming in *Publications of the Institute of British Geographers*.

Harman, H. H. (1960), *Modern Factor Analysis*. Chicago: The University of Chicago Press. Chapters 1–3, 9.

King, L. J. (1966), "Cross-Sectional Analysis of Canadian Urban Dimensions: 1951 and 1961." *Canadian Geographer*, Vol. 10, pp. 205–224.

Lawley, D. N., and Maxwell, A. E. (1963), *Factor Analysis as a Statistical Method*. London: Butterworth & Co. (Publishers), Ltd., Chapters 1, 4, 5.

Wong, S. T. (1963), "A Multi-variate Statistical Model for Predicting Mean Annual Flood in New England." *Annals*, Association of American Geographers, Vol. 53, pp. 298–311.

CLASSIFICATION AND
REGIONALIZATION PROBLEMS

In any area of science, classification is an important step in the ordering of knowledge and the deriving of generalizations. In geography, problems of classification appear in two different although related forms. The first is the general problem of grouping together observations or phenomena which exhibit certain levels of similarity in their characteristics. Such is the problem of classifying cities according to their employment structures or manufacturing industries on the basis of their locational preferences. The second form of classification is more peculiar to geographic research and relates to the identification or delimitation of *regions* which are uniform with respect to either a set of characteristics or an orientation to some urban center or node. Hence, geographers seek to identify agricultural and manufacturing regions and urban tributary regions.

These classification and regionalization problems now can be handled with some very powerful analytical techniques, the most important of which are discussed in this chapter.

8

8.1 DISTANCE STATISTICS AS MEASURES OF SIMILARITY

On a linear scale the similarity of observations or values can be measured in terms of the distances separating them either from one another or from a chosen point. As an example, this was the philosophy underlying Nelson's (1955) classification of United States cities.

In two-dimensional space, distances still serve as measures of similarity. If the space is a Cartesian one, that is, with orthogonal axes, then the distance between any two points is given by the Pythagorean theorem. This was noted in the earlier discussion of distance measures for point patterns and was illustrated in Figure 5.2. Assuming that the distances are to serve as measures of similarity, then it is essential that the two axes be measured on the same scale. Transformation of both variates into standardized Z scales satisfies this requirement.

If the two variates (X_1, X_2) defining the space also are correlated, then the effect of this correlation on the location of the observations in the two-space must be removed by a rotation of the axes. The rotation is according to the angle α, where $\cos \alpha = r_{12}$.

Once the axes are standardized and rotated, then the distances separating the observations in the two-space are measures of their similarities and can be used to group the observations. Berry (1960b), King (1962b), and Mayfield (1967) used this technique in classifying urban centers. The corresponding diagram from King's study is reproduced here as Figure 8.1.

When more than two variates are involved, orthogonalization can be achieved by a principal components analysis. Stone (1960) used this approach in his study of British economic regions. The original set of p variates is transformed into a new set of orthogonal components, and scores on these components are obtained for each of the observations. As we noted in the previous chapter, there will be as many components as there are variables, but often only the first few components are emphasized since these usually account for most of the original total variance. Assuming there are m components ($m \leqslant p$), then the distance for each pair of observations (A, B) is

$$D_{AB} = \left[\sum_{i=1}^{m} \sum_{j=1}^{m} (X_{iA} - X_{iB})(X_{jA} - X_{jB}) \right]^{1/2}$$
$$= [(X_A - X_B)^T \cdot (X_A - X_B)]^{1/2}, \quad \text{in vector form.}$$

In the formal statistics literature, distance statistics usually have been discussed with reference to groups or populations rather than independent observations as outlined above.

The outstanding work in this field is associated with the Indian statistician, P. C. Mahalanobis. Arising from his anthropometric studies of racial groups

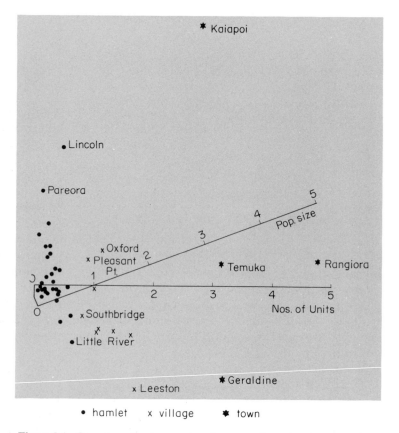

Figure 8.1. Grouping of urban centers in a two-dimensional space. The axes have been rotated to allow for the correlation between the two variates. On the ordinate one unit represents 731 persons; on the abscissa it represents 53 functional units. *Source:* King (1962b).

in Calcutta, Mahalanobis (1936) derived a generalized distance statistic D^2, defined as follows:

$$D_{ab}^2 = \sum_{i=1}^{p} \sum_{j=1}^{p} S^{ij} d_i d_j,$$

where

$$d_i = \bar{X}_{ia} - \bar{X}_{ib},$$ the difference in the two group means on variate i,

$$d_j = \bar{X}_{ja} - \bar{X}_{jb},$$ for $(i, j = 1, 2, \ldots, p,$

and

$$S^{ij} = (S_{ij})^{-1},$$ the element of the inverse of the common dispersion matrix.

The theory concerning the sampling distribution of the D^2 statistic assumes a p variate normal distribution. Rao (1965, p. 435) characterizes such a distribution by the property "that every linear function of the p-variables has a univariate normal distribution." This differs from the usual definition in terms of a probability density function (Anderson, 1958, pp. 5–44). The assumptions which follow, however, are the same in both cases.

There are two random samples (N_a, N_b) drawn independently from two p variate normal populations which have a common dispersion matrix $[S]$. This assumption that the dispersion matrices do not differ significantly can be tested using methods outlined by Anderson (1958, pp. 247–250) or Cooley and Lohnes (1962). If the assumption is valid, then the common dispersion matrix can be estimated by the expression

$$[S] = \frac{[S_a] + [S_b]}{(N_a + N_b - 2)},$$

where $[S_a]$ and $[S_b]$ are the sample sums of squares and cross products matrices in deviation form. For example,

$$[S_a] = \begin{bmatrix} x_{1a}^2 & x_{1a}x_{2a} & \cdots & x_{1a}x_{pa} \\ \cdot & & & \cdot \\ \cdot & & & \\ \cdot & & & \\ x_{pa}x_{1a} & & \cdots & x_{pa}^2 \end{bmatrix}.$$

In the case of only two groups, the D^2 statistic provides a test of the difference between the mean vectors of the two populations. The appropriate test statistic is given by Rao (1965, p. 480) as

$$F = \frac{N_a + N_b - p - 1}{p} \cdot \frac{N_a N_b}{(N_a + N_b)(N_a + N_b - 2)} \cdot D^2$$

with

$$\text{d.f.} = p, (N_a + N_b - p - 1).$$

Also, in the two-sample situation the D^2 statistic is closely related to another important multivariate statistic, T^2. The latter is used in testing hypotheses about the population mean vector and the significance of the difference between the mean vectors for two samples drawn from multivariate normal populations assumed to have a common but unknown dispersion matrix (Anderson, 1958, pp. 101–125). The relationship is

$$T^2 = \frac{N_a N_b}{N_a + N_b} \cdot D^2.$$

When more than two groups are involved, a matrix of D^2 statistics can be derived. King (1967) used such a matrix in analyzing the differentiation between urban growth categories in Ontario and Quebec. The results, shown in Table 8.1, suggested that on the variables considered the interprovincial

TABLE 8.1. D^2 STATISTICS FOR URBAN GROWTH CATEGORIES, CANADA 1951

Group	I	II	III	IV
I Quebec upward shift	—			
II Quebec downward shift	12.0	—		
III Ontario upward shift	32.4	47.2	—	
IV Ontario downward shift	37.7	45.9	1.1	—

Source: King (1967), p. 574.

contrasts were stronger than those between the growth categories. The work of Stone (1960, 1966) in analyzing the comparative economic structure of British regions has already been referred to in this chapter. In these studies, distance statistics were computed for all pairs of 12 economic regions and also for 4 groups of these regions.

8.2 CLASSIFICATION BASED ON DISTANCE MEASURES

Given a set of observations for which there is no existing classification, then the matrix of distance statistics for all pairs of observations can provide a basis for such a classification. The smaller the distance value, the more similar the observations in question are. Therefore, it should be possible in a series of discrete steps to group together observations which are close to one another. Ward (1961, 1963) has operationalized this procedure, and his algorithm has been applied particularly by Berry and his associates to a number of regionalization and classification problems in geography. Following Berry (1967a), the steps involved are described as follows:

(i)   The matrix of D^2 statistics for all pairs of observations is computed.

(ii)   The smallest D^2 is identified, and the two corresponding observations are grouped.

(iii)   The row and column vectors for these two observations in the distance matrix are replaced by a single row and column. These now contain the distances from the centroid of the new two-member group to all other observations.

(iv)   The process is continued through a series of steps until eventually only one group containing all the observations exists. This is in contrast to the start of the grouping when there were as many groups as observations.

At each step of the grouping it is possible to compute a value for an objective function, usually the pooled within-groups sum of squares. The selection of the groups to be joined at each step then is governed by the criterion that the increment to this pooled within sum of squares must be minimized. Berry (1967a) notes that the result is "a complete linkage tree," which "proceeds from n outermost branches, through $(n - 1)$ and $(n - 2)$ to i and $(i - 1)$ to 4, 3, 2 and finally the main trunk 1." No analytic solution exists for the problem of deciding how many groups are to be identified, although the step at which the ratio of the increment in the pooled within sum of squares to the total is at a minimum is suggested as a convenient point to terminate the grouping. As Berry observes, however, "it is possible to select that level deemed most desirable for a particular problem, and know that an optimal minimum-variance stratification has been achieved. . . ."

Berry has made extensive use of this grouping algorithm in his studies of urban retail structure and economic regionalization in eastern Canada; these were referenced in the previous chapter. Mention also has been made of the study by King (1966) concerned with the grouping of Canadian cities. The results in this case showed that the city groupings in Canada were remarkably stable over the period 1951–61, notwithstanding the fact that the important urban dimensions upon which the grouping was based changed quite considerably over the same decade.

There is no guarantee that the grouping procedure outlined above will yield contiguous regions when it is used on a set of areal units. The grouping is done solely with reference to the distance statistics, and regional types rather than contiguous regions are to be expected in most cases. It may be, however, that the contiguity effects and autocorrelation functions are such that the grouping does produce contiguous regions in some instances. For example, consider the maps in Figures 8.2 and 8.3. The first shows the agricultural regions in Ohio recognized by Weaver (1954) as part of his study of crop-combination regions in the Midwest. Compare this map with Figure 8.3, which shows the four regions given by the grouping algorithm above when it was applied to the component scores for the counties derived in the previous chapter. Not only are the two maps remarkably similar, but the grouping algorithm produced quite contiguous regions. Besides, it yields additional results which are not available in Weaver's study. In Figures 8.4 and 8.5, for instance, there are shown the groupings of counties for three regions and five regions, respectively, which are the optimal groupings for these different levels of regional aggregation.

In problems where contiguous regions are not assured but are desired, then a constraint can be placed on the grouping algorithm. The D_{AB}^2 statistics in the distance matrix are coded *positive* if observations A and B (areal units) are contiguous, *negative* if they are noncontiguous. The grouping algorithm then is constrained in that only positive distances are considered in the progressive building up of a group. Only in special circumstances would a whole row or column of the initial distance matrix be coded

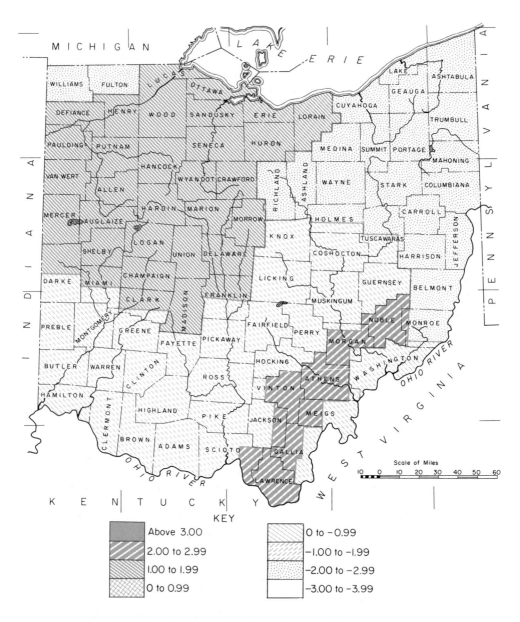

Figure 8.2. Weaver's crop-combination regions for Ohio. *Source:* Weaver (1954).

Figure 8.3. Crop-combination regions given by grouping algorithms. This map shows the same number of different regions as were recognized by Weaver.

Figure 8.4. Three crop-combination regions given by grouping algorithm.

Figure 8.5. Five crop-combination regions given by grouping algorithm.

negative. In the progressive manipulation of the distance matrix that takes account of established groupings, it is assumed that if an observation was contiguous to any one of the group members before the group was formed, it is contiguous to the group.

8.3 CLASSIFICATION AND LINEAR DISCRIMINANT FUNCTIONS

In this section we review some multivariate techniques which are useful for discriminating between populations or classifications and for assigning new observations to established classes with minimum probability of error.

As a statistical technique, *discriminatory analysis* stems largely from the work of statisticians who were concerned with biological and anthropometric data. In attempting to classify such phenomena as plants, animals, or even bone specimens, these research workers frequently encountered problems of indecision in that certain observations were noted which could conceivably belong to more than one class or group. Some means of discriminating among the given classes and of facilitating the classification of new cases with the least likelihood of error were desired. The development of discriminatory analysis was largely in response to these needs.

We begin by considering the simplest case of two *p* variate populations (or classes) of a similar kind, which overlap in the sense that certain members can be observed which might have arisen from either population. An obvious geographic analogy would be two core regions and an intermediate area which conceivably might be more similar to one of the core regions than to the other. How do we differentiate between the core regions, given the complexity of relationships among the important regional factors, and then how is the intermediate area to be assigned to one or other of the core regions?

The first solution is to derive a single linear discriminant function in the *p* variables so as to ensure not only maximum discrimination between the two populations but also minimum probability of error in assigning new individuals or objects to one or the other of the two populations. This analysis serves as an introduction to the more general solution involving more than two populations (or regions) and multiple linear discriminant functions, which are taken up in the second part of this section.

Two points should be emphasized before proceeding with the discussion. First, discriminatory analysis assumes that some form of classes are given. As Kendall (1957, p. 144) points out, "it is not the object of the inquiry to find what is the best way of dividing heterogeneous material into populations or classes." Nor does the analysis point up any inherent grouping tendencies in the data. Second, the use of the discriminant function in this context assumes that the individual or object to be classified belongs to one or the other of the two groups. Suspended judgments are not considered.

Beginning in 1936, with the publication of Fisher's papers, the problem of deriving discriminant functions as a means of differentiating between populations and facilitating the classification of new individuals or objects came to the fore. Fisher stated the problem as one of deriving a linear function of the observed variables $(X_1, X_2, \ldots, X_p)$,

$$Y = \lambda_1 X_1 + \lambda_2 X_2 + \cdots + \lambda_p X_p,$$

such that the ratio of the *difference between the sample means* to the *standard error within the two samples* would be maximized.

Given this requirement, then it can be shown that the solution for the coefficients in the linear discriminant function is given in matrix form as

$$[\lambda] = [S]^{-1} \cdot [d],$$

where the matrix $[S] = [S_a] + [S_b]$, that is, the pooled within sum of squares in deviation form which was discussed earlier, and the vector

$$[d] = [\bar{X}_{1a} - \bar{X}_{1b}, \bar{X}_{2a} - \bar{X}_{2b}, \ldots, \bar{X}_{pa} - X_{pb}]^T.$$

An alternative approach to this same problem makes use of a dummy dependent variable Y. Consider the case of N observations divided into two groups or samples of n_a and n_b members, respectively. Each observation in sample a is given a Y score of (n_b/N), whereas the members of sample b are given the value $(-n_a/N)$. These scores are arbitrary, but they ensure that the mean Y score is zero.

The discriminant function is now $Y = b_1 X_1 + b_2 X_2 + \cdots + b_p X_p$, which is, of course, the form of the multiple regression equation except that the constant is omitted. In classifying observations, this value would be of no importance since it is the same for all items. The values for the b coefficients are obtained by application of the least squares technique.

Having derived the discriminant function, a mean predicted Y score can be obtained for each group simply by inserting in the equation the values of $(\bar{X}_1, \bar{X}_2, \ldots, \bar{X}_p)$ for each group, respectively. With reference to the midpoint of the distance between these two mean values, new individuals then can be classified on the basis of their Y scores. The expected frequency of misclassification can be arrived at by use of a "t" statistic;

$$t = \frac{\frac{1}{2}(\bar{Y}_a - \bar{Y}_b)}{\left[\dfrac{n_a n_b \sum\limits_{}^{p} b_j d_j (1 - \sum\limits_{}^{p} b_j d_j)}{N(N - p - 1)} \right]^{1/2}}, \qquad \text{d.f.} = (N - p - 1).$$

The significance of the discrimination can be tested by an analysis of variance. The following sums of squares are required:

$$SS_{\bar{g}} = \frac{n_a n_b}{N} \cdot [\sum_{}^{p} b_j d_j]^2 \qquad \text{d.f.} = p,$$

$$SS_w = \frac{n_a n_b}{N} \cdot \sum^p b_j d_j [1 - \sum^p b_j d_j], \quad \text{d.f.} = (N - p - 1),$$

$$SS_T = \frac{n_a n_b}{N} \cdot \sum^p b_j d_j, \quad\quad\quad \text{d.f.} = (N - 1).$$

The test statistic is

$$F = \left(\frac{SS_g}{p}\right) \Big/ \left(\frac{SS_w}{N - p - 1}\right), \quad \text{d.f.} = p, (N - p - 1).$$

The ratio of the sum of squares for groups to the sum of squares for total provides an index of the discrimination R^2, the square root of which is in effect a biserial multiple correlation coefficient since the dependent variable is dichotomous. If the discrimination were perfect, that is, if $R = 1$, then this would mean that every predicted score for individuals in sample a would be (n_b/N) and for sample b, $(-n_a/N)$.

An example of this simple discriminatory analysis is given now. In Figure 8.6 a number of counties in eastern South Dakota are shown. These include two core regions as designated by the United States Department of Agriculture (1950). One of these regions, A, is an area of "specialized wheat farming," whereas the other, B, is a region of "feed grains and livestock." Located between these two areas are 10 counties which were not classified into either of the two core regions. Assuming that classification along these lines is desired, then the problem is handled easily within the framework of discriminatory analysis.

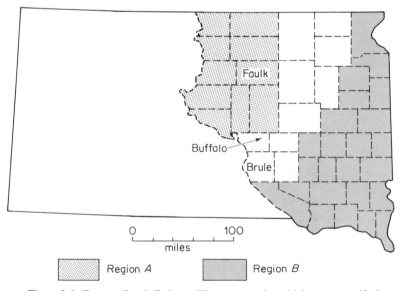

Figure 8.6. Eastern South Dakota. The ten counties which were classified by use of the linear discriminant function discussed in the text lie between the two regions.

For the counties, data were obtained on four selected variables. These variables were chosen simply on grounds of expediency in order to illustrate the technique under discussion. In a research context, there would be a great deal more attention given to the choice of relevant variables. The ones chosen were

  (i) X_1, density of rural farm population per square mile, 1950.

  (ii) X_2, normal annual precipitation in inches.

  (iii) X_3, percent of county area in flat land.

  (iv) X_4, average size of farm, 1950.

Through use of the dummy variate approach, the following linear discriminant function was obtained:

$$Y = 0.0393X_1 - 0.0811X_2 - 0.0065X_3 + 0.0922X_4.$$

The mean predicted score for Region A was -0.8065, whereas the corresponding figure for Region B was -1.5284. The midpoint between these two means was -1.1674, so that any new county having a Y value greater than -1.1674 was included in Region A, whereas counties with Y values smaller than -1.1674 were classified as belonging to Region B. Although the inferential questions were meaningless since no sampling was involved in this example, the analysis of variance used to test the significance of the discrimination is given in Table 8.2.

TABLE 8.2.   ANALYSIS OF VARIANCE FOR DISCRIMINATORY ANALYSIS

Sources	SS	d.f.	ms
Groups	3.6783	4	0.9195
Within	1.4167	29	0.0488
Total	5.0950	33	

$$F = 18.84, \quad \text{d.f.} = 4, 29,$$
$$F_{05} = 2.70, \quad F_{01} = 4.04.$$

The computed F exceeded the tabled F at both the 95 and the 99 percent levels of confidence, hence the conclusion that the discrimination was significant. The index of discrimination was 0.7219, whereas the obtained value of "t" was 1.6339. The probability of such a value being exceeded in repeated sampling lies between 10 and 20 percent. In other words, over the long run, in classifying 100 new counties, somewhere between 10 and 20 of them would be misclassified.

Of the 10 counties which were then classified by use of the discriminant function, only 2, namely Buffalo and Brule, were classified as belonging to Region A. The remaining 8 were included in Region B. Only 1 of the core counties appeared misclassified. Faulk County in Region A appeared to belong more properly to Region B.

Regionalization problems in geography typically involve more than two regions and, therefore, the analysis above should be generalized to the case of m regions. In addition, for a p variate situation, one linear discriminant function may not be sufficient to achieve maximum discrimination between the given groups or regions. The generalization of the analysis then also should allow for the extraction of more than one linear discriminant function. This is the problem we consider now.

Hodges (1950) and his colleagues in a series of reports have reviewed a number of general problems in discriminatory analysis. Anderson (1958), Cooley and Lohnes (1962), Kendall (1957), and Rao (1952, 1965) also provide excellent discussions of the analysis from different points of view. Mosteller and Wallace (1964) discuss many aspects of discriminatory analysis in the context of their provocative study of the disputed authorship of some of the Federalist papers.

In the geography literature, Casetti (1964a, b) has outlined a general solution to the problem of discriminating among several groups with more than one linear discriminant function. The solution is such that interclass variability is maximized in proportion to total variance. The problem of classification is handled conveniently in the same framework. Casetti's discussion differs somewhat from that given in more formal texts, such as Anderson (1958) and Rao (1952, 1965), which emphasize probability concepts, but the overall thrust of the analysis is in the same direction. It is Casetti's approach which is summarized below.

Define the original data matrix as $[X]$, which is an $N \times p$ matrix with elements X_{hij}, where

$h = (1, 2, \ldots, m)$, the number of classes,
$j = (1, 2, \ldots, p)$, the number of variates,
$i = (1, 2, \ldots, n_h)$, the number of observations in the hth class.

Then using *dot* notation, where the dot replaces the subscript over which there is a summation, define

$X_{hi.}$ A row vector $= (X_{hi1}, X_{hi2}, \ldots, {}_{hip})$.

$\bar{X}_{h.j}$ A scalar representing the mean of the jth variable for the hth class.

$\bar{X}_{..j}$ A scalar representing the mean of the jth variable for all N observations.

$[\bar{X}.]$ The matrix obtained from $[X]$ by substituting for every X_{hij} the appropriate average $\bar{X}_{h.j}$.

$[\bar{X}]$ The matrix obtained from $[X]$ by substituting for every X_{hij} the appropriate average $\bar{X}_{..j}$.

$[V]$ The matrix obtained from $[X]$ by substituting for every X_{hij} the deviation $(\bar{X}_{h.j} - \bar{X}_{..j})$. Thus $[V] = [\bar{X}.] - [\bar{X}]$.

Finally, define $[W]$, the matrix obtained from $[X]$, by substituting for every X_{hij} the deviation $(X_{hij} - \bar{X}_{..j})$. Thus $[W] = [X] - [\bar{X}]$.

Hence, in the original p variate problem space every observation is represented by a vector $X_{hi.}$, every class by a vector $\bar{X}_{h..} = (\bar{X}_{h.1}, \bar{X}_{h.2}, \ldots, \bar{X}_{h.p})$, and the overall classification by the vector $\bar{X}_{...} = (\bar{X}_{..1}, \bar{X}_{..2}, \ldots, \bar{X}_{..p})$.

We now seek some linear combinations of the variates $(X_1, \ldots, X_p)$ which will allow an optimal discrimination between the classes in a new discriminant space.

Associated with any one of these linear combinations, say the zth, define, y_{hiz} as the value on the zth linear discriminant function for the ith observation in the hth class. That is,

$$y_{hiz} = b_{z1}X_{hi1} + b_{z2}X_{hi2} + \ldots + b_{zp}X_{hip},$$

where the b's are the coefficients to be derived. Define a vector of these y_{hiz} scores as

$$[Y_z]^T = (y_{11z}, y_{12z}, \ldots; y_{21z}y_{22z}, \ldots; y_{m1z} \ldots).$$

Then

$$[Y_z] = [X] \cdot [B_z].$$

It follows that the mean y_{hiz} for the hth class is

$$\bar{y}_{h.z} = \frac{1}{n_h} \sum_i y_{hiz},$$

and the grand mean of all the y_{hiz} on the zth function is

$$\bar{y}_{..z} = \frac{1}{N} \sum_h \sum_i y_{hiz}$$

Define $[\bar{Y}_{.z}]$ as the vector obtained from $[Y_z]$ by substituting for every y_{hiz} the appropriate mean $\bar{y}_{h.z}$. Then it can be shown that

$$[\bar{Y}_{.z}] = [\bar{X}_{.}] \cdot [B_z].$$

Similarly, $[\bar{Y}_z]$, the vector obtained from $[Y_z]$ by substitution of $\bar{y}_{..z}$ for every y_{hiz}, equals $[\bar{X}] \cdot [B_z]$.

Casetti notes that the definitions and statements given above amount to a *discriminant procedure* whereby the original p variate space is collapsed by way of a linear transformation into a z variate discriminant space. In this discriminant space the observations and the classes again are represented as points. For example, the vector $[Y_{hi.}]$ identifies an observation; the vector $[\bar{Y}_{h..}]$, a class; and the vector $[\bar{Y}_{...}]$, the entire classification.

Discriminant procedures may differ according to

(i) The type of transformation used; in the present discussion it is assumed to be linear.

(ii) The rules for determining the class representative points; assumed here always to be the class centroid or mean vector.

(iii) The distance function used for measuring the relative location of points—in this case the ordinary distance formula for Euclidian space.

The solution for the linear discriminant functions demands that the following ratio R be maximized.

$$R = \frac{\frac{1}{N} \sum_h n_h (\bar{y}_{h.z} - \bar{y}_{..z})^2}{\frac{1}{N} \sum_h \sum_i (y_{hiz} - \bar{y}_{..z})^2}.$$

The numerator of this expression relates to the average distance from the points representing the group centroids to the overall center of gravity. The denominator gives the average distance from the points representing the observations to the overall center of gravity. Therefore, maximizing this ratio ensures that interclass variability will be as high as possible in comparison to total variability.

In terms of the vectors defined above,

$$R = \frac{(\bar{Y}_{.z} - \bar{Y}_z)^T \cdot (\bar{Y}_{.z} - \bar{Y}_z)}{(Y_z - \bar{Y}_z)^T \cdot (Y_z - \bar{Y}_z)}$$

$$= \frac{B_z^T V^T V B_z}{B_z^T W^T W B_z}.$$

The solution for $[B_z]$, which maximizes R, involves the characteristic equation

$$|G - \lambda I| = 0,$$

where

$$G = (W^T W)^{-1} V^T V \quad \text{and} \quad \lambda = \frac{B_z^T V^T V B_z}{B_z^T W^T W B_z}.$$

The solution of this equation gives p eigenvalues and the associated characteristic vectors. The elements of each vector are the coefficients of a linear discriminant function, and the associated eigenvalue is an absolute measure of the discriminant power of that function. The ratio $\lambda_z / \sum_r^p \lambda_r$ gives the proportion of interclass variability expressed by the function.

Casetti discusses five properties of linear discriminant functions which should be noted at this point.

(i) The characteristic vectors mentioned above are orthogonal.

(ii) The number of linearly independent discriminant functions is at most equal to the lesser value of $(m - 1)$ or p.

(iii) The vectors $[Y_z]$ will be orthogonal only if the original data were orthogonalized.

(iv) The coefficients in the discriminant functions and the observation scores on these functions can be used to give empirical interpretations to the linear discriminant functions. This is similar to the problem of interpreting the principal components.

(v) Some of the linear discriminant functions may be ignored in subsequent analysis because of their limited contributions to the discrimination. The distances between points in the discriminant space will be altered accordingly.

An example which Casetti presents of this type of analysis is summarized here. For the 19 regions of Italy grouped into north, center, and south, data were collected on 25 measures of demographic, sociological, and economic phenomena. As we noted in the previous chapter, these data were orthogonalized by a principal components analysis, and the scores on the first *six* components which accounted for 89 percent of the total variance were retained for the discriminant analysis. The results of the latter are shown in Tables 8.3 to 8.5.

TABLE 8.3. PROPORTION OF DISCRIMINANT POWER ACCOUNTED FOR BY FIRST TWO DISCRIMINANT FUNCTIONS; ITALIAN REGIONS

Linear discriminant function	Proportion of discriminant power	Cumulative percent
1	0.95	0.95
2	0.05	1.00

Source: Casetti (1964a, p. 35).

TABLE 8.4. LOADINGS OF VARIABLES (COMPONENTS) ON LINEAR DISCRIMINANT FUNCTIONS (LDF's)

Variable	LDF 1	LDF 2
1	0.60	0.07
2	−0.19	0.19
3	−0.59	0.11
4	0.19	0.78
5	−0.06	−0.27
6	0.45	−0.52

Source: As for Table 8.3.

The first linear discriminant function alone accounted for 95 percent of the intraclass variability, and this function was seen as measuring "the progressiveness, the overall social and economic level of development of the Italian regions." The scores (Table 8.5) supported this interpretation. No attempt was made to interpret the second function.

These results concerning the unidimensional basis of the classification of Italian regions into north, center, and south were seen by Casetti as con-

TABLE 8.5.   OBSERVATION SCORES ON LINEAR DISCRIMINANT
              FUNCTIONS

Observation	LDF 1	LDF 2
Region 1: north		
Lombardia	0.64	0.36
Piemonte	0.57	0.11
Val D'Aosta	0.60	0.56
Trentino	0.48	0.48
Veneto	0.32	0.00
Friuli Venezia Giulia	0.61	−0.39
Liguria	0.53	−0.08
Emilia Romagna	0.40	0.00
Region 2: center		
Toscana	0.29	−0.21
Umbria	0.17	−0.61
Marche	0.16	−0.28
Lazio	−0.11	−0.21
Abruzzi Molise	−0.31	−0.13
Region 3: south		
Campania	−0.59	0.18
Publie	−0.84	0.07
Basilicata	−0.70	−0.01
Calabria	−0.65	0.05
Sicilia	−0.68	0.30
Sardegna	−0.89	−0.13

Source: Casetti (1964a, p. 36).

firming "a widely accepted belief, which, however, is often criticized as being a sweeping generalization which does not take into consideration the presence of depressed areas in the North and of socially and economically developed areas in the Center and South" (p. 24–25).

8.4 OPTIMAL GROUPING AND REGIONALIZATION

The analysis in the preceding section assumed that the classification was given and fixed. In many problems, however, this initial classification or regionalization may not be an optimal one in which intraclass variability is minimized and interclass variability maximized. Therefore, it may be both desirable and possible to improve upon the classification by reassigning certain observations to different classes.

Casetti (1964b) has considered this problem also in the context of discriminatory analysis. His discussion again is summarized here.

Consider an initial classification of N observations into m classes. This is represented as

$$A_s = [a_{s,t_i,i}],$$

where s is the particular classification and i is an observation which in this classification is assigned to class t_i.

Now apply a discriminant procedure D to obtain a new classification. This step is written as

$$DA_s = [Da_{s,t_i,i}]$$
$$= A_{s+1}$$
$$= [a_{s+1,t_i,i}].$$

We now establish that two classifications are equal if

$$a_{s,t_i,i} = a_{s+1,t_i,i} \qquad \text{for all } i = 1, 2, \ldots, N.$$

Therefore, repeated applications of the discriminant procedure can be represented as a power of D. Hence,

$$DA_1 = A_2$$
$$DA_2 = A_3 = D(DA_1) = D^2 A_1$$

$$\cdot$$
$$\cdot$$
$$\cdot$$

$$DA_{N-1} = A_N = D^{N-1} A_1.$$

These series of steps are the *discriminant iterations.*

The question arises as to when a limit classification is reached. In a sequence of classifications

$$A_1, A_2, \ldots, A_v, \ldots, A_w, A_{w+1},$$

A_w is a limit classification when

$$A_z = A_w \qquad \text{for } z \geqslant w$$

and

$$A_z \neq A_w \qquad \text{for } z < w \ (w \text{ is a finite positive integer}).$$

Casetti proves that for an initial classification of a finite number of observations this limit will be reached in a finite number of discriminant iterations. The proof makes use of the result that the total within-classes variability of a classification generated by a discriminant iteration from an initial but nonlimit classification will be less than the corresponding variability of the initial classification. The relevant theorems are given in Casetti (1964b).

Casetti also discusses at length the stability of limit classifications. Many different initial classifications may result in the same limit classification, although the reverse is not true. An initial classification has but one limit classification. Depending upon the arrangement of the observations in the problem space, there will be a number of limit classifications of N objects into m classes. In Figure 8.7a, for example, there is one limit classification for the division into two classes. Contrast this with Figure 8.7b where there is an infinite number of limit classifications for division into two classes.

The stability of a limit classification, therefore, will be greater the larger the number of nonlimit classifications which generate it. For instance, no one limit classification in Figure 8.7b is very stable in this respect.

(a)

The actual discriminant iterations algorithm involves the following steps:

(i)   Multiple linear discriminant functions are derived for the initial classification, along with the scores of the observations on these functions.

(ii)   The distances of each observation from each class centroid in the discriminant space are calculated.

(iii)   For each observation, the shortest of these distances is identified.

(b)

(iv)   On this basis, observations are allocated to the nearest class, the data are reordered, and a new set of functions computed. The steps above are repeated until a limit classification is achieved.

Figure 8.7. Hypothetical distributions of observations in problem spaces for multiple discriminant analysis. After Casetti (1964b).

Casetti suggests that there are three major uses of discriminant iterations.

(i)   They can be used to *evaluate the relative quality of a classification*. There are a number of measures relevant in this context. First, how many iterations are required to generate a limit? Second, what are the differences between an initial classification and its limit or between different classifications at any level, with respect to the ratios

$$R^0 = \frac{\text{within variability}}{\text{total variability}} = 1 \Big/ \Big(1 + \sum_r^p \lambda_r\Big); \qquad 0 < R^0 < 1,$$

$$R = \frac{\text{between variability}}{\text{within variability}} = \sum_r^p \lambda_r; \qquad 0 < R < \infty?$$

If R^0 is small, then intraclass variability is relatively low and hence interclass differences must be high. By contrast, small values of R correspond to poor classifications, large values to good ones. Third, the number of observations in different classes in the limit classification as compared to the initial one is another measure of the quality of a classification.

(ii)   They serve to identify the cores of classifications and regionalizations. These comprise the observations which do not change their class from the initial to the limit classifications.

(iii)   They can be used in testing the validity of a classification with respect to sets of data chosen for different points in time or space. For example, an urban classification scheme devised for United States cities might be tested with respect to European data and evaluated in terms of some of the indices mentioned above.

SUGGESTED READINGS

Berry, B. J. L. (1961c), "A Method for Deriving Multi-Factor Uniform Regions." *Przeglad Geograficzny*, Vol. 33, pp. 263–282.

Berry, B. J. L. (1967a), "The Mathematics of Economic Regionalization." *Proceedings 4th General Meeting IGU Commission on Methods of Economic Regionalization*. Prague: Czechoslovak Academy of Sciences, pp. 77–106.

Kendall, M. G. (1957), *A Course in Multivariate Analysis*. London: Charles Griffin & Co., Ltd., pp. 111–116.

King, L. J. (1967), "Discriminatory Analysis of Urban Growth Patterns in Ontario and Quebec, 1951–1961." *Annals*, Association of American Geographers, Vol. 57, pp. 566–578.

Stone, R. (1960), "A Comparison of the Economic Structure of Regions Based on the Concept of Distance." *Journal of Regional Science*, Vol. 2, pp. 1–20.

EMERGING TRENDS
AND FUTURE PROSPECTS

9

In concluding this review of the applications of statistical analysis in geography, it is pertinent to comment on more recent developments along these lines and to speculate briefly on some of the possible avenues of future work. At least four developments and trends seem to warrant attention. The first is the least innovative and represents the extension of present work in multivariate analysis. In the following section of this chapter, for example, two recent applications of canonical correlation analysis in geographic research are reviewed. The future undoubtedly will see further applications of this type of analysis, along with different forms of factor analysis models and discriminatory analysis.

A second and more challenging development stems from an increased awareness of problems associated with the analysis of two-dimensional spatial series. The result has been that many of the techniques developed originally for time series analysis are being adapted now for the analysis of the two-dimensional series. The interest in this new line of work is not confined to geographers but rather to several of the earth sciences, and there is a growing literature on the application of two-dimensional Fourier analysis and spectral analysis, for example, to a wide range of data analysis problems. Some of the related technical issues are discussed briefly in Section 9.2.

Perhaps the most important development in contemporary geographic research is the greater emphasis being given to the development of theory and the structuring of related probability models. The development deserves mention in this book for two reasons. First, because most of the models involved are probability models and in many cases they are illustrative of Neyman's dynamic indeterministic approach, which was mentioned in the introduction to this text. Second, throughout the preceding chapters considerable emphasis has been given to the point that statistical analysis appears more meaningful when it is backed by theory. Recall that this was implicit in Curry's comments on Thomas's and Anderson's approach to the problem of modifiable units. Also, in discussing factor analysis, we noted that theories and related hypotheses are important in determining the choice of an appropriate model and solution. Therefore in geographic research a greater emphasis on theorizing and model-building promises to lay firmer foundations upon which statistical data analysis can be based. In this regard, the recent book by Chorley and Haggett (1967) marks a significant step forward. Selected applications of probability models in geographic problems are introduced in Section 9.3.

Finally, there are signs in geographic research today that in the future more importance will be placed on behavioral issues related to man's spatial activities. Harvey (1968) comments on this development. The psychology literature, from which this new geographic research draws much of its inspiration, contains references to some alternative mathematical and statistical models. Section 9.4 comments on some of these models and their implications.

9.1 APPLICATIONS OF CANONICAL CORRELATION ANALYSIS

In his study of Indian commodity flows, Berry (1966) has outlined a general field theory of a spatial system which relies upon the basic postulate that "the fundamental spatial patterns that summarize the characteristics of areas and the types of spatial behavior that are the essence of the interactions taking place among the areas are interdependent and basically isomorphic." When it is expressed symbolically in terms of multivariate analysis, the field theory has the following form:

(i) The existing state of the system is summarized in two data matrices. One is an *attribute matrix A*, and it has as many rows as there are areas, namely *n*. Each column represents a characteristic or variable. The second matrix is the *interaction matrix*, which has as many rows as there are pairs of areas, namely $(n^2 - n)$, and as many columns as there are forms of interaction being studied.

(ii) A principal axes factor analysis of the attribute matrix produces, as part of the output, a factor scores or *structure* matrix. This is an $n \times s$ matrix, where *s* is the number of factors. From this matrix it is possible to compute a new matrix Δ, with $(n^2 - n)$ rows and *s* columns, which contains the distances between each pair of areas on each factor.

(iii) A factor analysis of the interaction matrix yields a factor scores or *behavior* matrix B, with $(n^2 - n)$ rows and *b* columns, where *b* represents the number of dimensions extracted by the factor analysis.

(iv) Canonical correlation analysis of the two matrices Δ and *B* allows the statement of relationships between the two bases of the regional system, the structural or uniform on one hand and the behavioral or functional on the other.

It is the subject of this canonical correlation analysis that we wish to consider here. The formal mathematics of the technique are given in Hotelling (1936) and Anderson (1958), and the reader is referred to these discussions for an elaboration of the points which are summarized below. Cooley and Lohnes (1962) and Rao (1965) also discuss the technique.

Assume two sets of variates X_j and Y_k which have a joint probability distribution. As in so much of the multivariate analysis, the assumption of normality is important only if questions of sampling theory and inferential tests are to be developed in connection with the technique.

Now the problem is to find linear combinations of the two sets of variables, say $U_r = \alpha^{(r)} X_j$ and $V_r = \beta^{(r)} Y_k$, such that U_r and V_r have unit variances and maximum correlation with one another but no correlation with other pairs $U_1, V_1; U_2, V_2; \ldots ; U_{r-1}, V_{r-1}$.

The solution requires the variance-covariance matrices for the different sets of variables. These are shown as partitions of the supermatrix $\sum$ as follows:

$$\Sigma = \begin{bmatrix} \sum_{xx} & \sum_{xy} \\ \sum_{yx} & \sum_{yy} \end{bmatrix}.$$

The matrix equation

$$\begin{bmatrix} -\lambda \sum_{xx} & \sum_{xy} \\ \sum_{yx} & -\lambda \sum_{yy} \end{bmatrix} \cdot \begin{bmatrix} \alpha \\ \beta \end{bmatrix} = 0$$

can then be derived consistent with the requirements for *U* and *V* noted above. The solution of the determinantal equation

$$\begin{vmatrix} -\lambda \sum_{XX} & \sum_{XY} \\ \sum_{YX} & -\lambda \sum_{YY} \end{vmatrix} = 0$$

yields a set of $(j + k)$ eigenvalues which turn out to be the canonical correlations for each successive pair of U's and V's. However, assuming that $j \leqslant k$, that is, the number of variates in the set X_j is less than or equal to the number in the set Y_k, then only j of these roots or eigenvalues will be positive. The remaining $k - j$ of them will be zero.

The eigenvectors corresponding to the j positive roots are obtained from the matrix equation given above. These eigenvectors contain the sets of coefficients α and β which are used to form the different linear combinations of U's and V's.

Anderson (1958, p. 296) notes that it is possible to derive single equations for the sets of α's and β's. These are as follows:

$$[\textstyle\sum_{XY} \sum_{YY}^{-1} \sum_{YX} - \lambda^2 \sum_{XX}]\alpha = 0,$$

$$[\textstyle\sum_{YX} \sum_{XX}^{-1} \sum_{XY} - \lambda^2 \sum_{YY}]\beta = 0.$$

The solution of the determinantal equation in this case yields the values $(\lambda_1^2, \lambda_2^2, \ldots, \lambda_j^2)$. Both Rao (1965) and Cooley and Lohnes (1962) emphasize this approach. In the latter text, the solution also is given in terms of the correlation matrices rather than the variance-covariance matrices. The results, nevertheless, are identical to those given by the solution summarized above.

Cooley and Lohnes (1962, p. 37) outline Bartlett's test for the significance of canonical correlations. The test statistic is

$$\chi^2 = -[N - 0.5(j + k + 1)] \ln \Lambda$$

with jk degrees of freedom. The criterion Λ is defined as follows:

$$\Lambda = \prod_{i=1}^{j} (1 - \lambda_i) \qquad \text{for } j < k,$$

where the symbol $\prod$ stands for "the product of."

The null hypothesis at the first step is that the Y_k variables are unrelated to the X_j variables. If this null hypothesis is rejected, then the first canonical correlation is considered significant, its contribution to Λ is removed, and the test is applied to the remaining $(j - 1)$ correlations. At the second step,

$$\Lambda' = \prod_{i=2}^{j} (1 - \lambda_i)$$

and

$$\chi^2 = -[N - 0.5(j + k + 1)] \ln \Lambda', \qquad \text{d.f.} = (j - 1)(k - 1).$$

Cooley and Lohnes note that in general with r roots removed

$$\Lambda' = \prod_{i=r+1}^{j} (1 - \lambda_i),$$

and the corresponding chi-square statistic has $(j - r)(k - r)$ degrees of freedom.

Some of the results of Berry's study (1966) are given in Table 9.1. The "dyadic factors" are those resulting from the factor analysis of the interaction matrix, and the "structural factors" are those extracted from the attribute matrix. Berry gave the results for only four canonical variates, presumably

TABLE 9.1.   CANONICAL VARIATES IN DYADIC-STRUCTURAL COMPARISONS

Total set of data		Canonical vectors			
		I	II	III	IV
Dyadic factors*	(1)	0.64	0.05	0.23	0.25
	(2)	0.19	0.33	−0.68	0.18
	(3)	0.59	−0.14	0.42	−0.68
	(4)	0.38	0.57	0.53	0.45
	(5)	−0.21	0.72	0.10	−0.47
Structural factors†	(1)	0.41	0.32	0.65	−0.39
	(2)	0.26	−0.40	0.43	0.45
	(3)	−0.12	−0.53	−0.09	−0.55
	(4)	−0.37	−0.40	0.35	0.01
	(5)	−0.35	−0.32	0.65	0.50
	(6)	0.35	−0.22	0.23	0.21
	(9)	0.28	0.05	−0.12	−0.32
Canonical correlation		0.35	0.23	0.12	0.08
Chi-square		106.30	34.82	13.58	3.88
Degrees of freedom		24	15	8	3
Probability		0.001	0.01	0.10	0.25
Λ for set 0.805 (chi-square 279.62)					

*Viz., (1) Western and southern specialities moving north and east.
      (2) Western and northwestern products moving east and south.
      (3) Eastern specialities moving west and south.
      (4) Selected post-hinterland relationships.
      (5) Movements within the western region.
†Viz., (1) Urban-manufacturing specialization.
      (2) Rice-producing agricultures.
      (3) Longitudinal contrasts in western and eastern agricultural specialities.
      (4) Potential accessibility within India.
      (5) Maize-wheat-grain specialities of the northwest.
      (6) Metals production: iron, manganese, nonferrous.
      (9) Salt production.

Source: Berry (1966, p. 254).

those for which the correlations were significant. Along with the results of canonical correlation analysis of subsets of the structural matrix, these findings prompted the following general conclusions:

(i) "Commodity movements between trade blocks in the western and southern regions on the one hand and the eastern region on the other are greater, the greater the difference in the urban-manufacturing specialization and/or the metal ores production of the trade blocks between which the flow is taking place."

(ii) "Commodity flows both within the broader western, eastern and southern regions and more locally within the hinterlands of a greater number of ports, involve to a substantial extent intra-regional redistribution of the regions' specialities."

(iii) "There are also transfers of agricultural specialities between agricultural regions and urban-manufacturing specialities between urban regions."

Another geographic study in which canonical correlation analysis was employed is that of Gauthier (1966). In this study of the São Paulo region, Brazil, three variables which served as surrogates for urban economic development first were matched against five "structural dimensions of the highway accessibility surfaces." The first two canonical correlations were significant, and the relationships were interpreted in terms of

(i) Accessibility to the São Paulo Maior region and the associated increases in the value of manufacturing.

(ii) Increases in accessibility to the Pioneer region and a growth in urban population.

In an attempt to discover how the linear combinations of variables behaved over time, Gauthier did additional canonical correlation analyses of the following sets of variables:

(i) Increases on the eight variables, 1940–1950.

(ii) Increases on the eight variables, 1950–1960.

(iii) Increases on the three development variables, 1940–1950; increases on the accessibility measures, 1950–1960.

(iv) Increases on the development variables, 1950–1960; increases on the accessibility components, 1940–1950.

The hypothesis was that "if the relationships between nodal accessibility and urban growth have been balanced over time, the highest canonical correlations will be associated with combinations (i) and (ii). If the relationships have been unbalanced, the highest correlations will be associated with either combination (iii) or (iv)." The statistical findings supported the notion of unbalanced growth with the highest correlations showing up on the fourth combination of variables. The "lead" effects of transportation development were highlighted as a consequence.

The studies by Berry and Gauthier are provocative in that they do illustrate a new approach to the problems of studying the interrelationships between different components of settlement and transportation systems. The geographer's traditional concern for the interrelationships between the spatial distributions of human activities and the characteristics of the physical environment would seem to suggest that this line of multivariate analysis potentially has even wider applications in geographic research.

In concluding this section, it is well to emphasize that other forms of multivariate analysis also hold promise for future geographic applications. In Harris (1963), for example, some more advanced forms of factor analysis are discussed particularly as they are relevant to the problems of studying change. Similarly, in Cattell (1966) a number of other forms of multivariate analysis are reviewed which would seem to warrant consideration in the context of geographic research.

9.2   ADVANCED ANALYSIS OF SPATIAL
SERIES

In Chapter 6 reference was made to some problems associated with the correlation of spatial series, and the concept of spatial autocorrelation was introduced at that point. These earlier references and comments were deliberately oversimplified. In fact, the discussion of autocorrelation functions and a number of other closely related topics form part of a body of theory concerning the statistical analysis of series of events.

This body of statistical analysis and theory has been developed in the past almost exclusively with respect to events occurring over time. Bendat and Piersol (1966) and Cox and Lewis (1966) provide excellent overviews of the subject and reference the important mathematical contributions to the development of the theory. More recently, the application of the techniques and theory to spatial rather than temporal series has commanded increasing attention. In geography, this new orientation is represented in the recent work of Casetti (1966), Curry (1966b), Rayner (1966, 1967), and Tobler (1966a, b). The geological studies of Harbaugh and Preston (1965), Preston (1966), and Krumbein (1966) and the meteorological investigations of Pierson, et al. (1960) and Leese and Epstein (1963) are illustrative of the parallel work in other disciplines.

The statistical analysis of series of events involves more advanced mathematics than have been used up to this point in the text. No attempt is made, therefore, to develop the mathematics of the subject here. Only the major topics are summarized along the lines suggested by Rayner (1966).

Consider a series of events for a random variable Y, which occurs along a time or distance continuum. The employment levels in a city or region would serve as an illustration of a temporal series, whereas the depth of soil at points along a traverse would be an example of a one-dimensional spatial series. It is of interest to describe these series with reference to the fluctuations and oscillations which they exhibit over time or distance.

Unfortunately, the two examples mentioned above are not really appropriate ones with which to introduce this discussion. For the techniques which were developed originally for the analysis of series of events usually assume that the series are *periodic;* that is to say, the same pattern of events is assumed to be repeated both forward and backward in time or distance. Obviously, this would not be the case with the examples cited, and as we shall note shortly, other techniques are more appropriate in these cases. For the present we assume a periodic series. This implies also that there is no overall trend in the data.

The fitting of trigonometric polynomials is an accepted approach in analyzing periodic data. The Fourier series

$$y = \sum_{k=0}^{\infty} (a_k \sin k\theta + b_k \cos k\theta)$$

can be fitted to any such series of data. The terms in this expression are explained best with reference to Figure 9.1. The abscissa of this diagram is measured in *radians*, and the basic interval for the set of data is defined as 2π radians or θ_n. Within this basic interval there are assumed to be n equally spaced data points, the interval between them being $2\pi/n$ radians. Any one observation is identified by a subscript j, which has the range 0, 1, 2, $\dots$, $(n-1)$. Hence the jth point or observation is $(2\pi j)/n$.

The value of k gives the *frequency*, that is, the number of waves or cycles per basic interval. In Figure 9.1 the frequency is 2 cycles per basic interval. The coefficients a_k and b_k are defined as follows:

$$a_k = A_k \sin \Phi_k,$$
$$b_k = A_k \cos \Phi_k,$$

where A_k is the amplitude. It is easily verified that $A_k = (a_k^2 + b_k^2)^{1/2}$ and that Φ_k, which is the phase angle, equals $\tan^{-1}(a_k/b_k)$. This phase angle involves the location of the first crest in relation to the ordinate. The value Φ_k/k is called the *phase shift*, and in Figure 9.1 this is equal to $\pi/6$ with $\Phi_k = \pi/3$.

The actual fitting of Fourier series to a function $y = g(x)$ is called *harmonic* or *Fourier* analysis. Assuming that the function is periodic and is repeated from $-\infty$ to $+\infty$, then the observed set of data can be used to compute the amplitudes, phase angles, and frequencies and these, in turn, can be used in analysis and interpolation. The coefficients a_k and b_k are derived as follows:

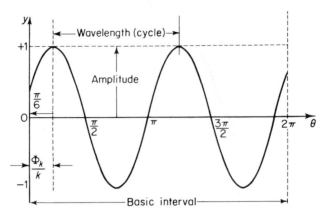

Figure 9.1. Example of a cosine curve. In this case, the frequency k is 2 cycles/basic interval, and the phase angle Φ_k is $\pi/3$. After Rayner (1966).

$$a_k = \frac{2}{n} \sum_{j=0}^{n-1} Y_j \sin k\theta_j$$

and

$$b_k = \frac{2}{n} \sum_{j=0}^{n-1} Y_j \cos k\theta_j.$$

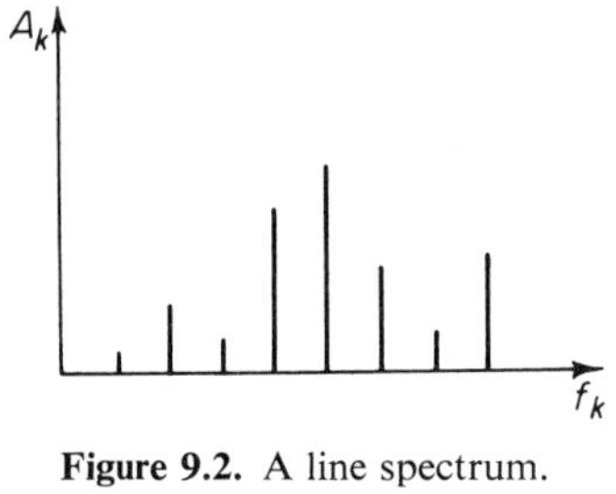

Then $A_k = (a_k^2 + b_k^2)^{1/2}$, and $\Phi_k = \tan^{-1}(a_k/b_k)$. The plot of the amplitudes A_k against the frequency k is referred to as the *line spectrum* (Figure 9.2).

The percentage contribution of different waves or cycles to the total variance can be computed as

$$(100)A_k^2 \Big/ \sum_{k=1}^{n/2} A_k^2.$$

Figure 9.2. A line spectrum.

The mathematical solutions in harmonic analysis may be derived either on the assumption that the function is truly periodic, in which case the integration for the a_k and b_k values is applied only in the basic interval $-\pi$ to $+\pi$, or on the assumption that the function is everywhere zero outside the interval for which the data are available. In the latter case the integration is done from $-\infty$ to $+\infty$, and the so-called Fourier transform is involved. These mathematical questions are considered in detail in books such as Lanczos (1956) and Tolstov (1962).

In the geography literature, the study of North American precipitation patterns by Horn and Bryson (1960) illustrates the use of Fourier analysis on a series which is assumed to be periodic. The studies by Lahey, et al. (1958); Peixoto, et al. (1964); Sabbagh and Bryson (1962); and Harbaugh and Preston (1965) provide other examples of interest to the geographer.

In the Harbaugh and Preston (1965) study, double Fourier series are introduced for two-dimensional series. Krumbein (1966) has reviewed the relative merits of this approach in comparison to the use of polynomial models in problems of geological interpolation and extrapolation.

In the two-dimensional context also, Casetti (1966) has discussed the significance of the size and shape of the areal units used in relation to the filtering out of different harmonics and the consequent effects on the correlations and regressions between two series.

In a sense, the Fourier analysis of periodic data is deterministic and there are no questions of statistical inference which are relevant. By contrast, *spectral analysis* of nonperiodic data, which is introduced at this point, does emphasize a number of inferential questions. Now the data are more properly regarded as random data and the set of observations as a sample set of possible outcomes. The assumptions which are now important are those of *stationarity* and *ergodicity*. Bendat and Piersol (1966, Chapter 1) give an excellent review of these topics. Stationarity implies that the moments (such as the mean) and the joint moments (such as the autocorrelation) of the series do not vary with time or distance, whichever defines the series. Ergodicity implies that the moments and joint moments do not vary from sample to sample. It is assumed, henceforth, that series satisfy both of these assumptions. Actually, if a series is ergodic, then it must be stationary.

The theory and computational techniques of spectral analysis are available in a number of excellent books, including Blackman and Tukey (1959), Lee (1960), Barber (1961), Cunnyngham (1963), Granger (1964), Hannan (1960), Bendat and Piersol (1966), and Cox and Lewis (1966), and in special issues of *Technometrics* (1961) and the *Journal of Applied Statistics* (1965).

Rayner (1967), in reviewing much of this work, notes that there are two approaches in the spectral analysis. One proceeds by way of computing the autocovariance function in the case of a single series and then applying a cosine transform to this function. There are a number of technical questions which are important in this context, namely the "filtering" and "prewhitening" of the data to remove particular frequencies, the application of a "window function" to allow for the fact that the autocovariances are computed from a finite sample, and the "aliasing" effects in the spectrum resulting from folding in frequencies which are not revealed by the data. These questions are summarized in Jenkins (1961). The spectrum of the autocovariance function which is obtained gives estimates of the variance in terms of frequency bands. Confidence intervals for the estimates of the spectrum are derived with reference to the χ^2 distribution. Tukey (1949) is the classic reference on this topic.

For two variables or series Y and Z, this first approach yields estimates of the *cospectrum* and the *quadrature spectrum*. The former is a sort of instantaneous covariance of the two series, whereas the latter is the out-of-phase covariance between the two series. From these estimates, it is possible to compute the cross amplitude $A_{yz}(f)$, the phase difference $\Phi_y(f)$ — $\Phi_z(f)$, and the coherence $W_{yz}(f)$. In these expressions, f is the central frequency of a band. The coherence, Rayner notes, varies between 0 and 1 and is "a measure of how well the two series are related in that band. This is a most useful measure of association between two sets of data. It supplies information on the degree of the association between data with respect to scale, something which is not available with the simple correlation coefficient. Furthermore, the phase difference provides estimates of the average direction of the relationship."

The alternative approach in spectral analysis is to calculate many values of the line spectrum and to average these over fairly wide frequency intervals. This approach does not require computation of the autocovariance function. Rayner notes that with the new emphasis on a computational routine known as the "fast Fourier transform" this approach in spectral analysis is being favored. Rayner (1967) himself discusses the use of this routine in outlining a spectral analysis approach to the problem of the correlation between two spatial series.

The application of spectral analysis to empirical geographic problems involving two-dimensional spatial series barely has commenced. Rayner (1967) discusses the necessary extensions of the mathematical analysis, as does Preston (1966). Tobler (1966b) suggests possible applications in social science analysis but stresses that the problems of data availability are formidable. Recall that the data points must be equally spaced and a minimum of 100–200 such points seems essential if reliable estimates of the spectrum are to be obtained.

Notwithstanding these data problems, the application of spectral analysis holds promise for some lines of future geographic research. Problems in physical geography provide obvious possibilities. Also, Bartlett (1963, 1964) has discussed the spectral analysis of point patterns, and this approach might prove useful, for example, in testing theories and models of spatial diffusion and even migration. Nor should it be overlooked that, as a form of data analysis, spectral methods might yield some useful inductive generalizations. Marschak (1966) makes a similar point about the application of the methods in econometrics, suggesting that "spectral analysis might help do away with the unrealistic time lags so cavalierly chosen by practicing econometricians." Geographers might have similar expectations with regard to the spectral analysis of spatial series.

9.3 PROBABILITY MODELS IN GEOGRAPHIC RESEARCH

It was suggested in the introduction to this chapter that the work involving probability models is one of the more important contemporary developments in quantitative geographic research. What are the bases for this contention? First is the fact that on a broader level the work is indicative of a concern for theorizing and structuring symbolic models about spatial distributions; although here we emphasize only probability models, it should not obscure the more general nature of this research thrust, and the current work on nonstochastic models is equally important in this context. The claim that geography is a science will appear all the more powerful when the discipline can boast of a body of theory concerning the spatial distributions of human and physical phenomena which is couched in the language of science and mathematics and which is supported not only by logic but by empirical findings.

A second point is related to the first one, and it has been made already in this text. The statistical analysis of data always appears more meaningful when it is backed by theory and when it involves the testing of hypotheses suggested by the theory. It is perhaps not surprising that so much of the recent quantitative work in the discipline has been in the field of urban geography, because the formulations of Christaller, Lösch, and Haig, for example, allowed for the deduction of many hypotheses which could be tested empirically. This work is far from completed, but already in urban and many other fields of geography there is a need felt for new and additional theoretical structures to provide the foundations for future deductive and inductive studies.

A third point concerns the emphasis upon probabilistic approaches. There is no easily developed rationale for this position. In books on the philosophy of science, for instance, Nagel (1961), the arguments are made that social science can aspire only to statistical and, hence, probabilistic theories in view of the nature of the phenomena with which it deals. On a more pragmatic level, Curry (1966c) notes that the probabilistic approach has the advantage that "it allows a problem to be approached with the explicit admission of considerable ignorance," although it may at the same time "lead to results which are intuitively inconceivable to our deterministically structured minds." Nevertheless, Curry's contention is that "the formal language of probability theory does hold the greatest hope of establishing a structure of explanation to which the various parts of geography can contribute and from which they can draw."

In supporting this argument, Curry presents a very good survey of the uses of probability theory in geographic research. His discussion is too lengthy to summarize here so the reader is referred to the paper in question. One or two of Curry's main points, however, are mentioned below.

One of the features of current geographic research which is reflected in the emphasis on probabilistic approaches is the heightened interest in questions of human behavior. In focusing on the individual's reaction to his physical, social, cultural, and economic environment, the geographer has found it appropriate to view the related decisions and actions in a probability framework. For example, Golledge (1967) has developed a model of marketing patterns in which the decision on the part of a farmer or producer to market or not to market his goods in particular urban centers involves an initial search process with different sets of outcomes and rewards. The "learning" process results in the eventual attainment of a pattern of stereotype behavior on the part of individual producers. The elapsed time before this state is reached is not independent of the location of the producer relative to the locations of competing market centers. Those located in the zones of competition will attain to the patterns of stereotype behavior later than those located in closer proximity to one of the major urban centers. The mathematical model involves a Markov process, which is described in such books as Bailey (1964) and Bharucha-Reid (1960).

Other geographic studies in which questions of individual behavior are assigned probabilities include Curry (1963, 1966a), Kates (1962), and Wolpert (1965). Closely related are the studies which make use of what Curry (1966c) describes as a "summation man," for which the choices or decisions made have a considerable random element. Such is the case with Marble's (1964) Markov chain analysis of urban travel patterns. The diffusion studies and migration studies associated with the work of Hägerstrand (1952, 1957, 1965, 1966, 1967), Kulldorff (1955), and Bowden (1965) are structured around basically the same concept of behavior, and so also are Curry's intriguing attempts (1962, 1966d) to formulate a new theory of central place systems in terms of queuing theory, initially, and, more recently, turbulence theory and communications theory.

On a far more aggregative level, probability models are being developed for intraurban and interurban systems. The work of Dacey, which is outstanding in this context, and of Curry (1964) has been referenced in earlier chapters of this text. Morrill (1962, 1963, 1965a) and Garrison (1962) have considered the problem of formulating probability models of urban systems developing through time. Much of this geographic work is being paralleled by developments in the field of urban and regional planning. Harris (1966), Steger and Douglas (1964), and Schlager (1964) provide examples of the latter work.

It is suggestive of the complexity of the situations with which these different studies deal that analytic solutions to the models seldom appear possible. In other words, the solutions can not be arrived at by formal mathematical analysis. The fact that we are dealing with probability models and, hence, probability functions does not, of course, preclude the possibility of such a solution. It might involve, for example, deriving a joint probability distribution for several variables. But in most geographic studies to date, and particularly those in which a time dimension is involved, the methods of *simulation* have been preferred to the more formal mathematical analysis.

Simulation as a technique encompasses a wide range of procedures, and references such as Geisler, et al. (1962); Galliher (1959); or Tocher (1963) provide discussions and illustrations of this point. The Monte Carlo method is by far the most widely used simulation technique. Hammersley and Handscomb (1964) discuss this method in detail. Hägerstrand (1965) led the way in applying it to problems of spatial analysis.

A Monte Carlo simulation involves random sampling from a known probability distribution function. For a one-dimensional series the procedure is illustrated in Figure 9.3. We have a sequence of random numbers from 0 to 1. One of these, say r_n, is chosen and this leads to the sampling of a value of x, namely x_n, as shown in the diagram. In two dimensions, the probability distribution function for the variable is mapped onto a set of grid squares,

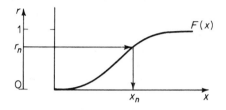

Figure 9.3. Illustration of Monte Carlo sampling. To the left of the graph of the distribution function is the random numbers scale. After Galliher (1959).

for example, the familiar 5 × 5 grid used by Hägerstrand (Figure 9.4). A random number between 0 and 1 is chosen and this identifies the particular grid square which on this trial is to be assigned a value of the variable. In the Hägerstrand studies, only a binary-coded variable is involved and the

0.0000–	0.0050–	0.0200–	0.0400–	0.0550–
0.0049	0.0199	0.0399	0.0549	0.0599
0.0600–	0.0750–	0.1050–	0.1575–	0.1875–
0.0749	0.1049	0.1574	0.1874	0.2024
0.2025–	0.2225–	0.2750–	0.7250–	0.7775–
0.2224	0.2749	0.7249	0.7774	0.7974
0.7975–	0.8125–	0.8425–	0.8950–	0.9250–
0.8124	0.8484	0.8949	0.9249	0.9399
0.9400	0.9450	0.9600	0.9800	0.9950
0.9449	0.9599	0.9799	0.9949	0.9999

Figure 9.4. Mapping of a probability distribution function on a two-dimensional grid for spatial simulation. After Hägerstrand (1953).

grid square receives a value of either 1 or 0 on any one trial. Pitts (1963), Taaffe, Garner and Yeates (1963), Yuill (1964), Morrill (1965b), Nystuen (1967), Törnqvist (1967), and Brown (1968) similarly have applied Monte Carlo simulation in the context of geographic research problems.

No attempt has been made in this section to discuss specific probability models. In part, this is because the range of possibilities is so wide and, in addition, much of the geographic work along these lines remains exploratory and undeveloped in very formal terms. The use of Markov chains and Markov processes, for instance, has been noted in the case of Marble's and Golledge's work, but these topics still remain foreign to most geographers. Gould (1966a) has referenced a number of "models of search" which have possible applications in geography and Brown (1965) has provided a similar review of models of diffusion, but the challenges represented in these compendiums have yet to be accepted. On a more basic level, the derivation of joint probability functions for spatial variables poses formidable mathematical problems, but at the same time it promises to yield many significant results. The development of this probability analysis in relation to patterns of locations and interactions as they vary over space and through time is a challenge for future geographic research. It should not go unanswered.

9.4 NON-EUCLIDEAN CONCEPTS AND THEIR IMPLICATIONS

In discussing the statistical techniques included in this book, the emphasis has been placed generally upon the algebraic expressions and interpretations. In most cases, however, there are particularly useful geometric interpretations which can be given to the same techniques. Reference is made again to Gould (1966b) for an illustration of how these geometric interpretations can aid in the understanding of quite complex multivariate problems.

The geometry which is relevant to most of the statistical techniques in this book is the Euclidean geometry. References to this field of mathematics are provided in Appendix A. One of the more important concepts in Euclidean geometry is that the distance between two points is given by the expression

$$d_{ij}^2 = (x_i - x_j)^2 + (y_i - y_j)^2 + (w_i - w_j)^2,$$

in the case of three-dimensional space. The extension to higher dimensions simply involves additional terms on the right.

In some of the recent psychological research on the problems of multidimensional scaling, the adequacy of this Euclidean distance measure as an index of the similarity of points in a space has been questioned. Alternative distance measures have been considered, specifically the "Manhattan" or "city-block" metric in which

$$d_{ij} = |x_i - x_j| + |y_i - y_j| + \cdots,$$

and the "Minkowski" metric whereby

$$d_{ij} = (|x_i - x_j|^r + |y_i - y_j|^r + \cdots)^{1/r}.$$

The "Manhattan" metric has been applied, for example, by Attneave (1950) and the Minkowski metric by Kruskal (1964). Torgeson (1958, 1965) provides an excellent discussion of scaling and the indexing of similarity and includes reference to some of the geometrical models relevant in these contexts. Guttman (1967), in discussing factor analysis solutions and their use in deriving metric information from nonmetric or ordinal data, also touches on the question of non-Euclidean geometries. The relevance of these studies to the problem of geographic space perception has been reviewed nicely by Downs (1967).

The subject of non-Euclidean geometries is a fairly wide one and an attempt is made to reference the more important topics in Appendix A. The term "non-Euclidean" should not be used loosely since it may refer to any one of a number of concepts in geometry. The findings of Luneberg (1947) in his study of binocular vision are non-Euclidean in the sense that they demand the concept of parallelism found in Lobachewskian rather than Euclidean geometry. Kruskal's use of the Minkowski metric is "non-Euclidean" however, in a more restricted sense. The basis for this generalization of the distance metric seems to lie more in Minkowski's work on number theory than in his four-dimensional geometry of space and time.

To date there have been no formal applications of non-Euclidean geometries in the geography literature, We might note that in Rushton, et al. (1967) there is a simple use of the Manhattan metric in measuring distances between consumer locations and urban places but, unfortunately, the authors attempt no discussion or justification of the method.

Projecting into the future, however, we might reasonably anticipate that the current research in a number of geography departments concerning human behavior over space and the human perception of spatial phenomena and environments will produce some published results along these lines. These studies probably will raise questions not only as to the appropriate geometrical models but also regarding the use of different measurement scales in geographic research. The two issues are not unrelated, as noted above in reference to the studies from the psychology literature.

Certainly, the use of multidimensional scales and data which are not of the interval or ratio type is suggestive of a wide range of new possibilities for the geographer interested in aspects of human behavior. One consequence of this development could be a far greater emphasis upon nonparametric and distribution-free techniques than has been the case in geography to date. This possibility, along with the others which have been hinted at throughout this text (for example, a greater emphasis on stochastic processes operating over time and space, the application of decision theory and Bayesian tech-

niques, the further development of multivariate analysis, and the consideration of alternative geometries and spatial transformations), offers an exciting
challenge for geographic research in the future.

SUGGESTED READINGS

Berry, B. J. L. (1966), "Essays on Commodity Flows and the Spatial Structure of
the Indian Economy." *Research Paper 111*, Department of Geography,
University of Chicago.

Casetti, E. (1966), "Analysis of Spatial Association by Trigonometric Polynomials."
The Canadian Geographer, Vol. 10, pp. 199–204.

Curry, L. (1967), "Quantitative Geography." *The Canadian Geographer*, Vol. 11,
pp. 265–279.

Dacey, M. F. (1966c), "A Probability Model for Central Place Locations." *Annals*,
Association of American Geographers, Vol. 56, pp. 550–568.

Garrison, W. L. (1962), "Toward a Simulation Model of Urban Growth and
Development." *Lund Studies in Geography, Series B*, Vol. 24, pp. 91–108.

Hägerstrand, T. (1967), "On Monte Carlo Simulation of Diffusion." In W. L.
Garrison and D. F. Marble (eds.), *Quantitative Geography*, Northwestern
University Studies in Geography, No. 13, pp. 1–32.

Rayner, J. N. (1967), "Correlation Between Surfaces by Spectral Methods." In
D. F. Merriam and N. C. Cocke (eds.), "Computer Applications in the Earth
Sciences: Colloquium on Trend Analysis." *Computer Contribution 12*, State
Geological Survey, University of Kansas, pp. 31–37.

Wolpert, J. (1965), "Behavioral Aspects of the Decision to Migrate." *Papers*,
Regional Science Association, Vol. 15, pp. 159–169.

APPENDIX A

SOME MATHEMATICAL GUIDELINES

The serious student of mathematical statistics and probability theory will require a much more formal training in mathematics than has been assumed for the readers of this book. Hopefully, the future will see a far greater number of geographers who have this formal training in mathematics, and for these persons this appendix has nothing to offer. But for other students of the discipline who are interested in improving a meager background in mathematics, this appendix attempts to plant some directional signs to the lines of mathematical analysis which appear relevant and important in the context of current geographic research.

A.1 INTRODUCTION TO MATHEMATICS

There are certain introductory topics such as number systems, functions, and equation systems which are basic to many lines of mathematical analysis, and these essentials must be mastered by those wishing to read further in mathematics.

These topics are reviewed in a nonrigorous manner in the books

W. W. Sawyer, *Mathematician's Delight* (Harmondsworth, England: Penguin Books Ltd., 1943).

———, *A Prelude to Mathematics* (Harmondsworth, England: Penguin Books Ltd., 1955).

H. M. Walker, *Mathematics Essential for Elementary Statistics* (New York: Holt, Rinehart and Winston, Inc., 1951).

These readings should be supplemented, however, by a review of the more formal treatments given in any college-level introductory mathematics text. There are advantages in using texts written especially for social scientists, and in this regard the following books are very useful:

R. G. D. Allen, *Mathematical Analysis for Economists* (London: Macmillan & Co., Ltd., 1956).

T. Yamane, *Mathematics for Economists* (Englewood Cliffs, N.J.: Prentice-Hall, Inc., 1962).

R. McGinnis, *Mathematical Foundations for Social Systems* (Indianapolis: The Bobbs-Merrill Co., Inc., 1965).

J. M. Shapiro and D. R. Whitney, *Elementary Analysis and Statistics: Mathematics for Administrative, Behavioral, and Biological Sciences* (Columbus, Ohio: Charles E. Merrill Books, Inc., 1967).

Reference might be made also to

W. Rudin, *Principles of Mathematical Analysis* (New York: McGraw-Hill Book Company, 1964).

A.2 GEOMETRIES

Geometries have a special appeal to geographers since these mathematics deal with points, lines, surfaces, and spaces—concepts which the geographer so often finds to be useful representations of the real-world phenomena in which he is interested. The close conceptual ties between geometries and certain lines of geographical analysis are stressed in the recent important methodological works of Bunge (1962) and Haggett (1965).

The classical geometry is that of Euclid. His systemization of the subject has proved to be the most enduring example in science of the formal axiomatic method whereby one begins with certain axioms and then deduces a set of consequences or theorems. In Euclid's work there are "definitions," "axioms," and "postulates," although in more recent statements of his geometry the distinction between axioms and postulates has been ignored. We shall note later that Euclid's postulates have been the subject of much subsequent theorizing and that investigations into one of these postulates in particular, the so-called fifth postulate or parallel postulate, prompted the discovery of some new non-Euclidean geometries.

The central theorem of Euclidean geometry is the Pythagorean theorem, which states that for any right-angled triangle in a plane the square on the hypotenuse is equal to the sum of the squares on the other two sides. Hence, in the following diagram,

$$AC^2 = AB^2 + BC^2.$$

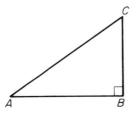

The student can review this theorem and its proof, along with the other theorems and postulates of Euclidean geometry, in an introductory text such as

H. G. Forder, *Geometry* (London: Hutchinson & Co., Ltd. 1950).

or in more advanced texts, for example,

H. G. Forder, *The Foundations of Euclidean Geometry* (New York: Dover Publications, Inc., 1958).

K. Borsuk and W. Szmielew, *Foundations of Geometry* (Amsterdam: North Holland Publishing Co., 1960).

An extremely important development in the history of mathematics was the formulation of *analytical geometry*. Attributed in origin largely to the work of the French scholar Descartes, analytical geometry involves the application of algebra to the study of geometry. The geometric properties of a space are studied with reference to a coordinate system and there is a one-to-one correspondence between points in the space and finite sequences of numbers. In the diagram below, for instance, the point P is fixed by the pair of numbers (x_p, y_p).

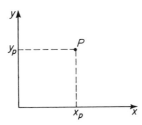

The set of coordinates are referred to often as Cartesian coordinates.

For an introduction to the subject of analytical geometry the reader is referred to

Allen, *op. cit.*, Chapter III.

Forder, *Geometry, op. cit.*, Chapter III.

Shapiro and Whitney, *op. cit.*, Chapter 6.

C. O. Oakley, *Analytic Geometry* (New York: Barnes & Noble, Inc., 1949).

The Pythagorean theorem makes it possible to derive an expression for the distance between two points in a Cartesian space. The distance between points P and Q in the following diagram is derived as follows:

$$PQ^2 = PS^2 + SQ^2$$
$$= (y_P - y_Q)^2 + (x_Q - x_P)^2;$$

therefore

$$PQ = \sqrt{(y_P - y_Q)^2 + (x_Q - x_P)^2}.$$

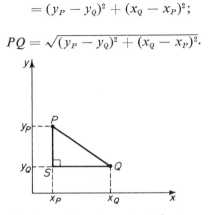

This distance formula is another important feature of Euclidean plane geometry. A similar expression can be derived for spaces of more than two dimensions; the spaces are still Euclidean and the only difference is that the formulae for the distance between the two points have additional terms on the right under the square root sign. We already have made use of such expressions in Chapter 8 on problems of classification.

The extension of Euclidean geometry to spaces of more than two dimensions is basic to the geometric interpretation of multivariate statistical analysis. Harman (1960, Chapter 4) and Gould (1966b) in their respective discussions of factor analysis outline some of the relevant extensions. A more formal discussion is provided in

> D. M. Y. Somerville, *An Introduction to the Geometry of N Dimensions* (New York: Dover Publications, Inc., 1958).

In these discussions of the geometry of higher dimensions there often is a need for an understanding of trigonometry. For example, Harman discusses how in a general N-dimensional Cartesian coordinate system, that is, one in which the coordinates are not necessarily at right angles to one another, the distance between two points P_1 and P_2 is given as

$$P_1P_2 = \sqrt{\sum_i \sum_k (x_{1i} - x_{2i})(x_{1k} - x_{2k}) \cos \theta_{ik}},$$

where i and k refer to the coordinates or dimensions and θ_{ik} is the angle between a pair of coordinates. The symbol "cos" is the trigonometric function called *cosine*. The same function appears in the geometric interpretation of the correlation coefficients, for as Harman (1960, p. 62) notes, "the coefficient of correlation between two variates (measured as deviates from their respective means) is the cosine of the angle between their vectors in N-space."

The simple trigonometric functions can be defined with respect to the angle θ in the following diagram.

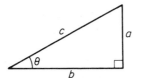

$$\text{sine of } \theta = \sin \theta = a/c$$
$$\text{cosine of } \theta = \cos \theta = b/c$$
$$\text{tangent of } \theta = \tan \theta = a/b$$
$$\text{cosecant of } \theta = \csc \theta = 1/(\sin \theta) = c/a$$
$$\text{secant of } \theta = \sec \theta = 1/(\cos \theta) = c/b$$
$$\text{cotangent of } \theta = \cot \theta = 1/(\tan \theta) = b/a$$

These functions are tabled for different angular values of θ. The mathematics of trigonometric functions are reviewed in

Shapiro and Whitney, *op. cit.*, pp. 166–180.

Trigonometric functions were used in this book in the sections on distance statistics and principal components analysis. Recall that in the simple two-variable classification problem the axes were rotated through an angle whose cosine equaled the correlation coefficient for the two variates in question. This rotation removed the effect of the correlation on the location of the observation points. In the discussion of principal components analysis, the concept of an orthogonal transformation was outlined initially in terms of trigonometric functions.

Reference was made earlier in this section to non-Euclidean geometries and to the fact that these were discovered largely as a result of formal consideration of the so-called parallel postulate of Euclid's system. The formulation of this postulate can be illustrated with reference to the diagram below.

In the Euclidean system, through the point P in the plane only one line PQ can be drawn parallel to the line RS. In this case, the two angles QPR and SRP add up to 180°. If, on the other hand, the sum of these two angles is less than 180°, then the two lines PQ and RS must meet somewhere to the right of the line PR; whereas if the sum is greater than 180°, the two lines must meet to the left of the line PR.

Considerable interest focused on whether or not this parallel postulate could be derived from the other axioms and postulates of Euclid's system. It was shown that it could not and that it was, in fact, quite independent of them.

The new geometries which were developed did not contain this parallel postulate. Independently, the two scholars Bolyai and Lobachewsky formulated a logically consistent geometry based on the axiom that through a point in a plane at least two distinct lines could be drawn parallel to another line in the plane. This geometry has come to be known as *hyperbolic geometry*. In Section 9.4 reference was made to some psychological research which is making use of this non-Euclidean geometry.

Spherical geometry as developed by the German mathematician Riemann and others does not contain any postulate of parallel lines. Further, the Euclidean notion of lines of infinite length is modified to allow lines to be of finite length and yet still unbounded. Such is the case with the great circle lines of a sphere. Coxeter (see below) has discussed the further modification of spherical geometry by Klein into what is known as *elliptical geometry*.

In the mathematics of geometry there are other concepts·in addition to parallelism which serve to characterize different geometries. With reference to some of these concepts, Coxeter (p. 19) has presented a very useful genealogy of the more important geometries. It is shown below.

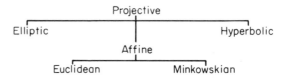

We have not discussed projective geometry, but it is one of the more basic areas of mathematics. It is concerned essentially with those graphical properties of figures which are preserved in any projection of the figures. In Sawyer's *Prelude to Mathematics* (Chapter 10) and Forder's *Geometry* (Chapter IV) there are good introductions to the subject. Affine geometry is Euclidean geometry except that the concept of "congruence" is not required. We say that two figures are "congruent" if they have the same shape and size, and "similar" if they have only the same shape. In affine geometry, Coxeter notes, there is not the concept of "perpendicularity" which is essential to Euclidean geometry.

Elliptic geometry contains the notion of congruence but similarity and parallelism are lacking. Hyperbolic geometry contains the concepts of congruence and parallelism (though different from Euclid's) but not similarity.

Minkowskian geometry has much in common with Euclidean geometry. Indeed, in his book, *Albert Einstein and the Cosmic Order* (New York: John Wiley & Sons, Inc., 1965, p. 9), Cornelius Lanczos notes that "Minkowski's view did not fundamentally alter our customary Euclidean geometry; it merely added one more dimension to it." This extra dimension was time,

and Minkowski developed a geometry in which space and time were insepar-
able. The formulation provided a geometrical translation for Einstein's
1905 "special relativity" theory. The later "general relativity" theory required
the spherical geometry formulated by Riemann. The books by Robb and
Rainich, which are cited below, outline some of the important differences
between Euclidean and Minkowskian geometries.

What is of interest in the behavioral research noted in Section 9.4 is
Minkowski's work on number theory and the geometry of numbers. The
so-called Minkowski inequality allows for the simple case

$$[(x_1 + x_2)^2 + (y_1 + y_2)^2 + (z_1 + z_2)^2]^{1/2}$$
$$< [x_1^2 + y_1^2 + z_1^2]^{1/2} + [x_2^2 + y_2^2 + z_2^2]^{1/2}.$$

Hardy, et al. (see below) point out that the geometrical interpretation of the
Minkowski inequality is the extension of this expression to a space of n
dimensions with a generalized definition of the distance between points given
as

$$P_1P_2 = (|x_1 - x_2|^r + |y_1 - y_2|^r + \dots)^{1/r} \qquad \text{for } r \geqslant 1.$$

It is this metric or distance which Kruskal (1964) used in his research on
multidimensional scaling. The mathematics of non-Euclidean geometries
and the number theory of Minkowski can be reviewed in these different
books.

Sawyer, *Prelude to Mathematics, op. cit.*, Chapter 6.

Forder, *Geometry, op. cit.*, Chapter V.

H. S. M. Coxeter, *Non-Euclidean Geometry* (Toronto: University of Toronto
Press, third edition, 1957).

G. H. Hardy, J. E. Littlewood, and G. Pōlya, *Inequalities* (Cambridge, U. K.:
The University Press, second edition, 1952).

G. Y. Rainich, *Mathematics of Relativity* (New York, John Wiley & Sons,
Inc., 1950).

A. Robb, *Geometry of Time and Space* (Cambridge, U. K.: The University
Press, 1936).

A.3　CALCULUS

We introduce now one of the more important and central areas of mathematics, namely calculus. The topics which are encompassed by this area of
mathematics are basic to many other lines of mathematical analysis, including
probability theory and statistics.

Calculus in itself is a wide subject for study. An introductory overview
with some good programmed instruction is provided in

D. Kleppner and N. Ramsey, *Quick Calculus* (New York: John Wiley &
Sons, Inc., 1965).

The books by Allen (1956), Rudin (1964), Shapiro and Whitney (1967), and Yamane (1962) which were referenced earlier in this appendix give lengthier and slightly more formal presentations of the subject. Among the numerous mathematical texts on calculus, the one following can be recommended:

> R. Courant and F. John, *Introduction to Calculus and Analysis* (New York: Interscience Publishers, 1965).

The study of calculus is the study of mathematical functions. A regression equation and a probability law are examples of mathematical functions to which calculus can be applied.

It is convenient to divide the subject matter of calculus into two parts, differential calculus and integral calculus. There is a theorem, known as the "fundamental theorem of calculus," which concerns the reciprocity between the two branches and which shows that differentiation and integration of a mathematical function are simply the inverse of each other. It is well to keep this unity in mind, even though at the introductory level it may appear that the topics can be studied separately.

The differential calculus studies the rate of change in mathematical functions. Given a function $y = f(x)$, it is of interest to know what the rate of change is in y for a very small incremental change in x. To study this question, we must have a firm foundation in the different forms and representations of mathematical functions, including the concepts of analytical geometry and trigonometry. Above all, we must be familiar with the notion of a *limit*, for this is central to calculus.

The mathematical definiton of a limit is provided in any one of the books already mentioned and it would be inappropriate to attempt to state this definition here. In crude terms, we may consider the limit of a function $y = f(x)$ to be that value of y which is approached as the value of x tends toward some particular value. This would be written as

$$\lim_{x \to a} y = \lim_{x \to a} f(x) = L.$$

Hence, for the function $y = x^2$,

$$\lim_{x \to 2} y = 4.$$

From the definition of a limit, a number of other important definitions follow, including that of *continuity*. A function $f(x)$ is said to be continuous at a point x_0 of the domain of the function, if the limit exists.

To define the derivative of a continuous function $y = f(x)$, it is convenient to introduce the symbols Δy and Δx to represent increments of y and x, respectively. Then

$$\lim_{\Delta x \to 0} \frac{\Delta y}{\Delta x} = \frac{dy}{dx}$$

is called the "derivative of y with respect to x." In a nonrigorous sense we may interpret this derivative as a sort of instantaneous rate of change for the function.

There are set rules for obtaining the derivatives of different types of functions and these rules can be found in any of the reference books already cited.

The derivative in a formal sense is a limit and is not to be regarded as a ratio of two amounts of change. The *differential* of a function $y = f(x)$ does tell, however, how much change we might expect in y for a small change in x. It is written as

$$dy = \frac{dy}{dx} dx.$$

Again it is worth stressing that the derivative dy/dx is regarded correctly as one symbol and not as a ratio. In applied problems the differential is extremely useful. For example, if we postulated a simple gravity model in which interaction (y) was an inverse function of the square of distance, say

$$y = k/r^2,$$

where k is a constant, then for a small change in distance we would have the following change in y:

$$dy = \frac{dy}{dr} dr.$$

$$= \left(\frac{-2k}{r^3}\right) dr.$$

Up to this point we have been considering only a single variable. Differentiation of functions involving several variables introduces *partial derivatives*. For instance, given the function

$$y = x^2 + 6xz + z^2,$$

we might be interested in the rate of change in y with respect to x while z was held constant. In this case, the partial derivative of y with respect to x is written as $\partial y/\partial x$, and it equals $2x + 6z$. Similarly, the partial derivative of y with respect to z is

$$\frac{\partial y}{\partial z} = 6x + 2z.$$

We shall note shortly that, in the regression analysis, taking partial derivatives is a step in obtaining the normal equations. First we need to consider one other topic in differential calculus, however, that of solving for the minimum and maximum values of a function. This is an important use of the differential calculus.

If a function has a minimum or maximum value for some value of x, then its derivative at that point will be zero, For example, in the following diagram, the derivative will be zero at the points x_0 and x_1.

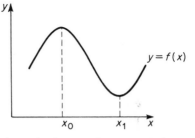

In order to determine whether we have a maximum or a minimum value of the function at some point at which the derivative is zero, assuming we do not have a graph of the function before us, we do in fact take the derivative of the derivative. In other words, we take the second derivative of the original function, which is written d^2y/dx. It can be shown that if $d^2y/dx < 0$, then $f(x)$ has a maximum value; if $d^2y/dx > 0$, then $f(x)$ has a minimum value. In the cases when the second derivative is zero, other tests must be used.

In regression analysis, the least squares method involves finding estimates of the constant a and the regression coefficients such that the sum of the residuals is minimized. Recall that in the two variable problem we had (p. 121),

$$U = \sum_{i=1}^{N} (X_{0i} - a - bX_{1i})^2,$$

which was to be minimized.

Taking the partial derivatives of U with respect to a and b gives

$$\frac{\partial U}{\partial a} = \sum_{i=1}^{N} [2(X_{0i} - a - bX_{1i})(-1)],$$

$$\frac{\partial U}{\partial b} = \sum_{i=1}^{N} [2X_{0i} - a - bX_{1i})(-X_{1i})].$$

Now if we set these equal to zero, divide through by the constant 2, and distribute the summation sign, we obtain

$$-\sum_{i=1}^{N} X_{0i} + \sum_{i=1}^{N} a + \sum_{i=1}^{N} bX_{1i} = 0,$$

$$-\sum_{i=1}^{N} X_{0i}X_{1i} + \sum_{i=1}^{N} aX_{1i} + \sum_{i=1}^{N} bX_{1i}^2 = 0.$$

Transposing and applying rules concerning the summation operator gives the normal equations

$$Na + b\sum_{i=1}^{N} X_{1i} = \sum_{i=1}^{N} X_{0i},$$

$$a\sum_{i=1}^{N} X_{1i} + b\sum_{i=1}^{N} X_{1i}^2 = \sum_{i=1}^{N} X_{0i}X_{1i}.$$

The same procedure holds in the multiple regression case except that there are more partial derivatives to be taken.

The problem of deriving maximum and minimum values has many interesting extensions. One involves the case of a function $y = f(x, z)$, where x and z in turn are connected by a function. Recall that this was the case in the solution of the principal components problem where we sought to maximize the variance associated with each component, in turn, subject to the constraint that the original correlations had to be reproduced. This was discussed on p. 172 of the text. A solution to this type of problem involves the Lagrange multiplier method; this method is discussed in

Allen, *op. cit.*, pp. 364–369.
Yamane, *op. cit.*, pp. 116–120.

We shall not attempt to describe the method here but simply illustrate it with an example taken from Yamane (pp. 116–118).

Assume a function $u = f(x, y) = x^2 - y^2 + xy + 5x$ and a subsidiary condition $g(x, y) = x - 2y = 0$. We wish to find the extreme value of u subject to this condition. The method involves forming the expression

$$z = f(x, y) + \lambda g(x, y),$$

where λ is an undetermined multiplier. Then we find

$$\frac{\partial z}{\partial x} = 0, \qquad \frac{\partial z}{\partial y} = 0, \qquad \frac{\partial z}{\partial \lambda} = 0$$

and solve these equations. For the example above, we have

$$z = (x^2 - y^2 + xy + 5x) + \lambda(x - 2y),$$

and

$$\frac{\partial z}{\partial x} = 2x + y + 5 + \lambda = 0,$$

$$\frac{\partial z}{\partial y} = x - 2y - 2\lambda = 0,$$

$$\frac{\partial z}{\partial \lambda} = x - 2y = 0.$$

Solving these equations gives the values

$$x = -2, \qquad y = -1, \qquad \lambda = 0, \qquad u = -5.$$

The method can be generalized to problems involving more than two variables and more than one constraint. These extensions are not presented here, but they are discussed in the sections of Allen and Yamane noted above.

We turn now to consider the other main branch of calculus, the *integral calculus*. This is no less important than the differential calculus; indeed, as we noted earlier, one is the inverse of the other.

The concept of an "integral," in turn, can be approached from two points of view. One involves the so-called "indefinite integral" which is obtained when we reverse the process of differentiation; that is to say, given a derivative, we obtain the corresponding function. Stated in another way, we are given a function and the object is to find a second function which has the first as its derivative. The second function, if it exists, is the indefinite integral.

For a function $f(x)$, the indefinite integral is written as $\int f(x)\,dx$. The process of deriving such integrals is part of the general process of *integration*. Some examples of indefinite integrals are given below:

$$\int x^2\,dx = (\tfrac{1}{3})x^3,$$

$$\int (x^2 + 3x + 2)\,dx = (\tfrac{1}{3})x^3 + (\tfrac{3}{2})x^2 + 2x,$$

$$\int \frac{1}{x}\,dx = \log x + \text{constant},$$

$$\int e^x\,dx = e^x + \text{constant}.$$

Integration is viewed also as the problem of finding the area under the curve of a function. For example, in the diagram below we could integrate to find the shaded area under the curve $y = g(x)$.

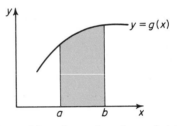

If we approach this problem from the viewpoint that the shaded area in question, as shown in the diagram below, is the sum of n strips each of width Δx and area $g(x')\,\Delta x$, where x' is some value of x in the interval Δx, then we can define the "definite integral" as the limit of this sum as Δx approaches zero.

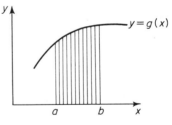

That is to say, the definite integral is

$$\lim_{\Delta x \to 0} \sum_{i=1}^{n} g(x')\,\Delta x = \int_a^b g(x')\,dx.$$

The fundamental theorem of calculus relates the two viewpoints concerning an integral and establishes that the definite integral of the function $g(x)$

between two points a and b is simply the value of the indefinite integral of $g(x)$ evaluated at b minus its evaluation at a. Hence,

$$\int_0^4 x^2 \, dx = (\tfrac{1}{3}) \cdot x^3]_0^4,$$

which is to say that the indefinite integral in question is $\tfrac{1}{3} \cdot x^3$ and that this has to be evaluated first for $x = 4$ and second for $x = 0$; then the difference between the two resulting values is the value of the definite integral $\int_0^4 x^2 \, dx$. In the present example, we have

$$\int_0^4 x^2 \, dx = (\tfrac{1}{3}) \cdot (4)^3 - (\tfrac{1}{3}) \cdot (0)^3$$
$$= (\tfrac{1}{3}) \cdot (64)$$
$$= 21.33.$$

It is not always possible to evaluate the indefinite integrals in question, and there are certain approximating techniques which have to be used. These are discussed along with the general rules for integration in the books on calculus referred to earlier.

The integral calculus has obvious importance in the study of continuous probability laws. Given a probability density function, then we can define the probabilities of values falling within certain limits as areas under the curve and obtain these by integration. The different published tables for continuous probability functions such as the normal, gamma, chi-square, "t", "F", and exponential distributions have been derived in this way.

The integral calculus also can be extended to the study of solids and used in obtaining volumes. The use of the so-called "demand cone" in the location theory of Lösch (1954, pp. 106–107) and others assumes that the total demand over the market area can be regarded as the volume of a right circular cone and this volume is given as an integral.

In this section a very elementary introduction has been given to some of the basic topics in calculus. As noted earlier, the serious student of quantitative research will have to study these topics in formal mathematics courses. Given the current trends in much social science research, including geography, toward the structuring of probabilistic models, then it is certain that a knowledge of at least intermediate calculus will be increasingly essential in the training of research geographers.

A.4 MATRIX ALGEBRA

As noted in the text, many of the problems in statistical analysis, particularly multivariate analysis, are handled most conveniently by using matrix algebra. This branch of mathematics is introduced in the general texts by McGinnis

(1965) and Yamane (1962) which have been referred to already. Other introductory statements and reviews of the subject are available in

> R. G. D. Allen, *Mathematical Economics* (London: Macmillan & Co., Ltd., 1959), Chapters. 12–14.
>
> T. W. Anderson, *Introduction to Multivariate Analysis* (New York: John Wiley & Sons, Inc., 1958), Appendix 1.
>
> H. H. Harman, *Modern Factor Analysis* (Chicago: The University of Chicago Press, 1960), Chapter 3.
>
> G. Tintner, *Econometrics* (New York: John Wiley & Sons, Inc., 1952), Appendices A.1, A.2.

More formal mathematical presentations of matrix algebra are presented in

> R. Bellman, *Introduction to Matrix Analysis* (New York: McGraw-Hill Book Company, 1960).
>
> F. E. Hohn, *Elementary Matrix Algebra* (New York: The Macmillan Co., 1964).
>
> M. Marcus and H. Minc, *Introduction to Linear Algebra* (New York: The Macmillan Co., 1965).

The following definitions are essential for a discussion of the methods of matrix algebra.

A *matrix* is a rectangular array of numbers which in a general sense are referred to as the elements of the matrix. A matrix may be represented symbolically as $[A]$ or $[a_{ij}]$, the latter form expressing the fact that there are i rows and j columns in the matrix. Each element is identified by a double subscript representing its row and column position.

A *square matrix* has as many rows as it has columns. This number is referred to as the *order* of the matrix. Thus a matrix $[A]$ of order 3 would be written as

$$\begin{bmatrix} a_{11} & a_{12} & a_{13} \\ a_{21} & a_{22} & a_{23} \\ a_{31} & a_{32} & a_{33} \end{bmatrix}.$$

A *diagonal matrix* is a square matrix with zero elements everywhere except on the principal diagonal. An example would be

$$\begin{bmatrix} 12 & 0 & 0 \\ 0 & 6 & 0 \\ 0 & 0 & 4 \end{bmatrix}.$$

A special case of a diagonal matrix is the *identity matrix* $[I]$, with ones on the principal diagonal. This is equivalent to the number *one* in matrix algebra. A *null matrix* has all zero elements. A *transposed matrix* is one whose rows and columns have been interchanged. In this text, the transpose of $[A]$ is written as $[A]^T$. Hence,

$$[A]^T = \begin{bmatrix} a_{11} & a_{21} & a_{31} \\ a_{12} & a_{22} & a_{32} \\ a_{13} & a_{23} & a_{33} \end{bmatrix}.$$

Two matrices are *equal* if they are of the same order and have equal corresponding elements. A *triangular matrix* is a square matrix in which all the elements either above or below the principal diagonal are zero. Two examples of triangular matrices are

$$\begin{bmatrix} 1 & 0 & 0 \\ 3 & 4 & 0 \\ 2 & 9 & 5 \end{bmatrix} \quad \text{and} \quad \begin{bmatrix} 3 & 6 & 13 \\ 0 & 2 & 3 \\ 0 & 0 & 1 \end{bmatrix}.$$

A *symmetric matrix* $[A]$ is one for which $[A] = [A]^T$. An example is given below.

$$\begin{bmatrix} 4 & 9 & 7 \\ 9 & 16 & 3 \\ 7 & 3 & 12 \end{bmatrix}.$$

The matrix of correlation coefficients is a symmetric matrix.

Matrix *addition* is defined not only for square matrices of the same order but also for any two matrices with the same numbers of rows and columns. The matrix sum is formed by adding the corresponding elements. For instance,

$$\begin{bmatrix} 3 & 6 & 2 \\ 4 & 9 & 7 \end{bmatrix} + \begin{bmatrix} 10 & 3 & 2 \\ 1 & 4 & 6 \end{bmatrix} = \begin{bmatrix} 13 & 9 & 4 \\ 5 & 13 & 13 \end{bmatrix}.$$

The *multiplication* of two matrices $[A]$ and $[B]$ is defined only in the case when the number of columns in $[A]$ equals the number of rows in $[B]$. Two square matrices of the same order can be multiplied and so also can the following pairs of matrices:

$$\begin{bmatrix} 3 & 6 & 4 \\ 2 & 6 & 1 \end{bmatrix} \cdot \begin{bmatrix} 6 & 1 & 9 & 2 \\ 3 & 8 & 4 & 1 \\ 6 & 9 & 3 & 6 \end{bmatrix}$$

$$\begin{bmatrix} 6 & 2 \\ 4 & 1 \end{bmatrix} \cdot \begin{bmatrix} 6 & 2 & 1 \\ 3 & 9 & 4 \end{bmatrix}$$

$$\begin{bmatrix} 3 & 2 & 4 & 2 \\ 9 & 6 & 7 & 3 \\ 12 & 2 & 10 & 1 \end{bmatrix} \cdot \begin{bmatrix} 1 & 3 \\ 6 & 2 \\ 7 & 9 \\ 10 & 4 \end{bmatrix}.$$

The general rule for multiplication is that it is "row times column" with the paired elements being multiplied and then the products summed. Hence,

$$\begin{bmatrix} a_{11} & a_{12} \\ a_{21} & a_{22} \end{bmatrix} \cdot \begin{bmatrix} b_{11} & b_{12} \\ b_{21} & b_{22} \end{bmatrix} = \begin{bmatrix} a_{11}b_{11} + a_{12}b_{21}, & a_{11}b_{12} + a_{12}b_{22} \\ a_{21}b_{11} + a_{22}b_{21}, & a_{21}b_{12} + a_{22}b_{22} \end{bmatrix}.$$

If we designate the elements of the product matrix on the right as

$$\begin{bmatrix} c_{11} & c_{12} \\ c_{21} & c_{22} \end{bmatrix},$$

then we can see that

$$c_{ij} = \sum_{k=1}^{2} a_{ik}b_{kj},$$

where

$$i = 1, 2 \quad \text{and} \quad j = 1, 2.$$

In the general case when $[A]$ is an $m \times r$ matrix and $[B]$ is an $r \times n$ matrix, then

$$c_{ij} = \sum_{k=1}^{r} a_{ik}b_{kj}, \quad \begin{aligned} i &= 1, \ldots, m \\ j &= 1, \ldots, n. \end{aligned}$$

Note that the product matrix is always an $m \times n$ matrix in this general case. We note also that, in general, matrix multiplication is not *commutative;* that is to say, $[A] \cdot [B]$ does not equal $[B] \cdot [A]$. This is illustrated below.

$$\begin{array}{ccc} A & \cdot \quad B & = \quad C \end{array}$$

$$\begin{bmatrix} 1 & 3 \\ 2 & 4 \end{bmatrix} \cdot \begin{bmatrix} 4 & 6 \\ 1 & 7 \end{bmatrix} = \begin{bmatrix} 7 & 27 \\ 12 & 40 \end{bmatrix}$$

$$\begin{array}{ccc} B & \cdot \quad A & = \quad D \end{array}$$

$$\begin{bmatrix} 4 & 6 \\ 1 & 7 \end{bmatrix} \cdot \begin{bmatrix} 1 & 3 \\ 2 & 4 \end{bmatrix} = \begin{bmatrix} 16 & 36 \\ 15 & 31 \end{bmatrix}.$$

One of the special cases in which the commutative rule does hold is for the *inverse matrix*. This is a particularly important concept in matrix algebra, for (as we shall see) it fulfills some of the roles of the operation of division which is not otherwise defined in matrix algebra.

If for a square matrix $[A]$ it is possible to find another matrix $[B]$ such that

$$[A] \cdot [B] = [I],$$

then

$$[A] \cdot [B] = [B] \cdot [A],$$

and $[B]$ is known as the inverse of $[A]$. In this text, the inverse matrix is written as $[A]^{-1}$. Hence,

$$[A] \cdot [A]^{-1} = [A]^{-1} \cdot [A] = [I].$$

The inverse matrix is extremely important in the solution of sets of equations. Recall that in the multiple regression problem the normal equations in deviation form were given as

$$[P_0] = [P_j] \cdot [B],$$

where $[P_0]$ was an $(m - 1) \times 1$ vector, $[P_j]$ was a square matrix of order $(m - 1)$, and $[B]$ was also an $(m - 1) \times 1$ vector. Then the solution for $[B]$ was obtained as

$$[P_j]^{-1} \cdot [P_0] = [B].$$

There are a number of different ways to solve for the inverse of a matrix. These are summarized nicely in the book

V. N. Faddeeva, *Computational Methods of Linear Algebra* (New York: Dover Publications, Inc., 1959).

Three of the methods are outlined below. The first involves the use of *determinants*, and in discussing this method, a number of other definitions concerning matrices will be established.

A *determinant* is a single value associated with a square matrix. In the case of a matrix $[A]$ of order 2, the determinant, written as $|A|$, is equal to $a_{11}a_{22} - a_{21}a_{12}$. That is,

$$|A| = \begin{vmatrix} a_{11} & a_{12} \\ a_{21} & a_{22} \end{vmatrix} = a_{11}a_{22} - a_{21}a_{12}.$$

In the case of higher-order matrices the determinant must be evaluated either by an expansion of the basic case noted above or by other procedures such as the "sweep-out" process, which is discussed in

Hohn, *op. cit.*, pp. 61–63.

For the expansion method it is necessary to define a *minor* and a *cofactor*. The minor M of an element a_{ij} in a square matrix is the determinant of the submatrix formed by deleting from the original matrix the ith row and the jth column. For example, given the matrix

$$\begin{bmatrix} 1 & 0 & 1 \\ 0 & 2 & 1 \\ 1 & 1 & 4 \end{bmatrix},$$

then

$$M_{11} = \begin{vmatrix} 2 & 1 \\ 1 & 4 \end{vmatrix} = 7,$$

$$M_{12} = \begin{vmatrix} 0 & 1 \\ 1 & 4 \end{vmatrix} = -1,$$

$$M_{13} = \begin{vmatrix} 0 & 2 \\ 1 & 1 \end{vmatrix} = -2,$$

and so on.

The *principal minors* of $[A]$ are those minors which are symmetric around the principal diagonal of $[A]$. If all the principal minors are positive and nonzero, then the matrix [A] is *positive definite*. If the principal minors are either positive or zero, then the matrix [A] is *positive semidefinite*. A matrix which is symmetric and positive semidefinite is referred to as a *Gramian* matrix. These properties of matrices are important in the solution of the characteristic equation associated with any square matrix. This topic will be considered later in this section. For the moment, we continue with the discussion of determinants and inverses.

A *cofactor* C_{ij} is the corresponding minor with the sign adjusted by $(-1)^{i+j}$. In the example above,

$$C_{11} = (-1)^2 M_{11} = 7,$$
$$C_{12} = (-1)^3 M_{12} = 1,$$
$$C_{13} = (-1)^4 M_{13} = -2.$$

Now the determinant can be evaluated by taking any one row or column and summing the products of the elements times their cofactors. For example, expanding along the first row, we have

$$\begin{vmatrix} 1 & 0 & 1 \\ 0 & 2 & 1 \\ 1 & 1 & 4 \end{vmatrix} = 1(7) + 0(1) + 1(-2),$$
$$= 5.$$

Alternatively, expanding down the third column gives the same result:

$$\begin{vmatrix} 1 & 0 & 1 \\ 0 & 2 & 1 \\ 1 & 1 & 4 \end{vmatrix} = 1(-2) + 1(-1) + 4(2),$$
$$= 5.$$

Some properties of determinants are worth noting. First, if any row or column of a matrix contains all zero elements, then the determinant is zero. The same result obtains if any two rows or two columns are equal or are simply multiples of each other. For example,

$$\begin{vmatrix} 1 & 0 & 1 \\ 0 & 2 & 1 \\ 1 & 0 & 1 \end{vmatrix} = 1(2) + 0 + 1(-2) = 0$$

and

$$\begin{vmatrix} 1 & 0 & 1 \\ 0 & 2 & 1 \\ 2 & 0 & 2 \end{vmatrix} = 1(4) + 0 + 1(-4) = 0.$$

The value of a determinant is in no way altered by an interchanging of corresponding rows and columns. Hence, $|A^T| = |A|$. If any two rows or columns of [A] are interchanged, however, then the determinant of the new matrix [B] is equal to $-[A]$. For example,

$$
\begin{matrix}
B & & A
\end{matrix}
$$

$$
\begin{vmatrix} 1 & 1 & 0 \\ 0 & 1 & 2 \\ 1 & 4 & 1 \end{vmatrix} = 1(-7) + 1(2) + 0 = -5 = -\begin{vmatrix} 1 & 0 & 1 \\ 0 & 2 & 1 \\ 1 & 1 & 4 \end{vmatrix}.
$$

The determinant remains unchanged if to any one row or column there is added some multiple of the elements of another row or column. For instance, if we take the matrix [A] given above and add to row 1 four times row 3, then we have

$$
\begin{vmatrix} 5 & 4 & 17 \\ 0 & 2 & 1 \\ 1 & 1 & 4 \end{vmatrix}.
$$

The determinant of this matrix is

$$
5(7) + 4(1) + 17(-2) = 5 = |A|.
$$

Finally, we note that the *rank* of a matrix [A] is the order of the largest submatrix in [A] which has a nonzero determinant. A square matrix which has a nonzero determinant is said to be *nonsingular,* and the rows and columns are linearly independent of one another. If the determinant equals zero, then the matrix is *singular* and contains either some rows or columns which are linearly dependent.

Gould (1966b) gives an interesting geometric interpretation of determinants.

Given a nonsingular matrix [A], we can find the inverse by use of the determinant and the matrix of cofactors. First, we write this matrix of cofactors for an order 3 matrix [A] as

$$
[J] = \begin{bmatrix} c_{11} & c_{12} & c_{13} \\ c_{21} & c_{22} & c_{23} \\ c_{31} & c_{32} & c_{33} \end{bmatrix}
$$

and transpose this to form what is called the *adjoint of A.* Thus

$$
[J]^T = [\text{adj } A] = \begin{bmatrix} c_{11} & c_{21} & c_{31} \\ c_{12} & c_{22} & c_{32} \\ c_{13} & c_{23} & c_{33} \end{bmatrix}.
$$

Then it can be shown that

$$
[A] \cdot [\text{adj } A] = \begin{bmatrix} |A| & 0 & 0 \\ 0 & |A| & 0 \\ 0 & 0 & |A| \end{bmatrix} = |A| \cdot [I].
$$

Now if we multiply each side by $[A]^{-1}$, we have

$$[A]^{-1} \cdot [A] \cdot [\text{adj } A] = [A]^{-1} \cdot |A| \cdot [I].$$

By definition $[A]^{-1} \cdot [A] = [I]$, and the product of any matrix times the identity matrix must equal simply the matrix itself. Therefore, the equation above reduces to

$$[\text{adj } A] = [A]^{-1} \cdot |A|.$$

Dividing each side by the single value $|A|$ gives

$$\frac{1}{|A|} \cdot [\text{adj } A] = [A]^{-1}.$$

This solution for the inverse is illustrated for the matrix given above:

$$[A] = \begin{bmatrix} 1 & 0 & 1 \\ 0 & 2 & 1 \\ 1 & 1 & 4 \end{bmatrix}$$

$$|A| = 5.$$

$$[J] = \begin{bmatrix} 7 & 1 & -2 \\ 1 & -3 & -1 \\ -2 & -1 & 2 \end{bmatrix}$$

$$[\text{adj } A] = \begin{bmatrix} 7 & 1 & -2 \\ 1 & 3 & -1 \\ -2 & -1 & 2 \end{bmatrix}$$

Therefore,

$$[A]^{-1} = \begin{bmatrix} \frac{7}{5} & \frac{1}{5} & -\frac{2}{5} \\ \frac{1}{5} & \frac{3}{5} & -\frac{1}{5} \\ -\frac{2}{5} & -\frac{1}{5} & \frac{2}{5} \end{bmatrix}.$$

As a check on the solution, $[A] \cdot [A]^{-1}$ should equal the identity matrix, which is the case.

$$\begin{bmatrix} 1 & 0 & 1 \\ 0 & 2 & 1 \\ 1 & 1 & 4 \end{bmatrix} \cdot \begin{bmatrix} \frac{7}{5} & \frac{1}{5} & -\frac{2}{5} \\ \frac{1}{5} & \frac{3}{5} & -\frac{1}{5} \\ -\frac{2}{5} & -\frac{1}{5} & \frac{2}{5} \end{bmatrix} = \begin{bmatrix} 1 & 0 & 0 \\ 0 & 1 & 0 \\ 0 & 0 & 1 \end{bmatrix}.$$

An alternative solution for the inverse involves what Hohn (pp. 103–105) calls the method of *synthetic elimination*. The matrix $[A]$ is first augmented on the right by an identity matrix of the same order as $[A]$. Then a series of operations are performed on $[A]$ to reduce it to an identity matrix. It can be shown then that the same sequence of operations performed on the identity matrix yields $[A]^{-1}$. Again the method is illustrated for the matrix $[A]$ given above:

$$\begin{bmatrix} 1 & 0 & 1 & \vdots & 1 & 0 & 0 \\ 0 & 2 & 1 & \vdots & 0 & 1 & 0 \\ 1 & 1 & 4 & \vdots & 0 & 0 & 1 \end{bmatrix}.$$

The steps are as follows:

(a) Subtract row 3 from row 2:

$$\begin{bmatrix} 1 & 0 & 1 & \vdots & 1 & 0 & 0 \\ -1 & 1 & -3 & \vdots & 0 & 1 & -1 \\ 1 & 1 & 4 & \vdots & 0 & 0 & 1 \end{bmatrix}.$$

(b) Add row 2 to row 3:

$$\begin{bmatrix} 1 & 0 & 1 & \vdots & 1 & 0 & 0 \\ -1 & 1 & -3 & \vdots & 0 & 1 & -1 \\ 0 & 2 & 1 & \vdots & 0 & 1 & 0 \end{bmatrix}.$$

(c) Add row 1 to row 2:

$$\begin{bmatrix} 1 & 0 & 1 & \vdots & 1 & 0 & 0 \\ 0 & 1 & -2 & \vdots & 1 & 1 & -1 \\ 0 & 2 & 1 & \vdots & 0 & 1 & 0 \end{bmatrix}.$$

(d) Subtract, from row 3, twice row 2:

$$\begin{bmatrix} 1 & 0 & 1 & \vdots & 1 & 0 & 0 \\ 0 & 1 & -2 & \vdots & 1 & 1 & -1 \\ 0 & 0 & 5 & \vdots & -2 & -1 & 2 \end{bmatrix}.$$

(e) Divide row 3 by 5:

$$\begin{bmatrix} 1 & 0 & 1 & \vdots & 1 & 0 & 0 \\ 0 & 1 & -2 & \vdots & 1 & 1 & -1 \\ 0 & 0 & 1 & \vdots & -\frac{2}{5} & -\frac{1}{5} & \frac{2}{5} \end{bmatrix}.$$

(f) Subtract row 3 from row 1:

$$\begin{bmatrix} 1 & 0 & 0 & \vdots & \frac{7}{5} & \frac{1}{5} & -\frac{2}{5} \\ 0 & 1 & -2 & \vdots & 1 & 1 & -1 \\ 0 & 0 & 1 & \vdots & -\frac{2}{5} & -\frac{1}{5} & \frac{2}{5} \end{bmatrix}.$$

(g) Add to row 2 twice row 3:

$$\begin{bmatrix} 1 & 0 & 0 & \vdots & \frac{7}{5} & \frac{1}{5} & -\frac{2}{5} \\ 0 & 1 & 0 & \vdots & \frac{1}{5} & \frac{3}{5} & -\frac{1}{5} \\ 0 & 0 & 1 & \vdots & -\frac{2}{5} & -\frac{1}{5} & \frac{2}{5} \end{bmatrix}.$$

The inverse of $[A]$ is now on the right.

Finally, in discussing the computation of the inverse, the *square root method* is outlined. The theory behind this method is given in

P. S. Dwyer, "The Square Root Method and its Use in Correlation and Regression." *Journal of the American Statistical Association*, Vol. 40, 1945, pp. 493–503.

Again the matrix [*A*] is augmented on the right by an identity matrix of the same order. Then corresponding to this identity matrix the square root method obtains a triangular matrix [S_m] such that

$$[S_m]^T \cdot [S_m] = [A]^{-1}.$$

For a square matrix of order 3 the method begins with

$$\begin{bmatrix} a_{11} & a_{12} & a_{13} & 1 & 0 & 0 \\ a_{21} & a_{22} & a_{23} & 0 & 1 & 0 \\ a_{31} & a_{32} & a_{33} & 0 & 0 & 1 \end{bmatrix}.$$

Call the elements of this augmented matrix p_{ij}. Then a matrix [*S*], of which [S_m] is a partition, is obtained as follows:

$S_{11} = \sqrt{p_{11}}$ $\qquad$ $S_{12} = p_{12}/S_{11}$ $\qquad$ $S_{13} = p_{13}/S_{11}$

$S_{14} = p_{14}/S_{11}$ $\qquad$ $S_{15} = p_{15}/S_{11}$ $\qquad$ $S_{16} = p_{16}/S_{11}$

$S_{21} = S_{12}$ $\qquad$ $S_{22} = \sqrt{p_{22} - S_{12}^2}$ $\qquad$ $S_{23} = (p_{23} - S_{12}S_{13})/S_{22}$

$S_{24} = (p_{24} - S_{12}S_{14})/S_{22}$ $\qquad$ $S_{25} = (p_{25} - S_{12}S_{15})/S_{22}$

$S_{26} = (p_{26} - S_{12}S_{16})/S_{22}$

$S_{31} = S_{13}$ $\qquad$ $S_{32} = S_{23}$ $\qquad$ $S_{33} = \sqrt{p_{33} - S_{13}^2 - S_{23}^2}$

$S_{34} = (p_{34} - S_{13}S_{14} - S_{23}S_{24})/S_{33}$

$S_{35} = (p_{35} - S_{13}S_{35} - S_{23}S_{25})/S_{33}$

$S_{36} = (p_{36} - S_{13}S_{16} - S_{23}S_{26})/S_{33}$

We have as a result

$$S_m$$

$$[S] = \begin{bmatrix} S_{11} & S_{12} & S_{13} & S_{14} & S_{15} & S_{16} \\ S_{21} & S_{22} & S_{23} & S_{24} & S_{25} & S_{26} \\ S_{31} & S_{32} & S_{33} & S_{34} & S_{35} & S_{36} \end{bmatrix},$$

and, as noted above,

$$[A]^{-1} = [S_m]^T \cdot [S_m].$$

In the example used above we would have

$$\begin{bmatrix} 1 & 0 & 1 & 1 & 0 & 0 \\ 0 & 2 & 1 & 0 & 1 & 0 \\ 1 & 1 & 4 & 0 & 0 & 1 \end{bmatrix}$$

$S_{11} = 1,$ $S_{12} = 0,$ $S_{13} = 1,$ $S_{14} = 1,$ $S_{15} = 0,$ $S_{16} = 0.$

$S_{22} = \sqrt{2} = 1.41421,$ $S_{23} = 1/1.41421 = 0.70711,$

$S_{24} = 0,$ $S_{25} = 1/1.41421 = 0.70711,$ $S_{26} = 0.$

$S_{33} = \sqrt{(4 - 1 - 0.5)} = 1.58114,$

$S_{34} = -1/1.58114 = -0.63246,$

$S_{35} = -0.5/1.58114 = -0.31623,$

$S_{36} = 1/1.58114 = 0.63246.$

Thus

$$[S_m] = \begin{bmatrix} 1 & 0 & 0 \\ 0 & 0.70711 & 0 \\ -0.63246 & -0.31623 & 0.63246 \end{bmatrix}.$$

Premultiplying this by its own transpose gives

$$[A]^{-1} = \begin{bmatrix} 1.4 & 0.2 & -0.4 \\ 0.2 & 0.6 & -0.2 \\ -0.4 & -0.2 & 0.4 \end{bmatrix},$$

which agrees with the solution given by the other two methods.

Reference was made earlier to the characteristic equation associated with a square matrix. For such a matrix $[A]$ of order n, this equation is

$$|A - \lambda I| = 0,$$

where I is the identity matrix and λ is a scalar.

This type of equation appears, for example, in problems involving the solution of a set of homogeneous linear equations, that is to say, equations in which all the variable terms are of degree one and the constants are zero. This was the case, recall, in the general solution for the principal components as outlined by Harman and reviewed on p. 173 of this text. There were p homogeneous linear equations involving the correlations as the coefficients and the component loadings as the variables. This set of equations in matrix form was as follows:

$$[R - \lambda I] \cdot [w_{j1}] = 0, \qquad \text{for } j = 1, \ldots, \text{p.}$$

It can be proved that in order for a set of equations of this form to have a nontrivial solution, that is, one for which not all the w_{j1} are zero, the coefficient matrix must be singular. Hohn (pp. 136–140) discusses this general topic. In other words, for the present example, it must be the case that $|R - \lambda I| = 0$, otherwise the solution for the w_{j1} is trivial.

The solution of the general characteristic equation $|A - \lambda I| = 0$ yields n values of λ, assuming that the matrix $[A]$ is of order n. This solution can be illustrated for the simple case when n equals 2. Given the matrix

$$[A] = \begin{bmatrix} 2 & 1 \\ 2 & 3 \end{bmatrix},$$

then

$$|A - \lambda I| = \begin{bmatrix} 2 - \lambda & 1 \\ 2 & 3 - \lambda \end{bmatrix} = \lambda^2 - 5\lambda + 4.$$

This equation is set equal to zero and solved for λ. The solution of any quadratic $ax^2 + bx + c$ is given by the expression

$$x = \frac{-b \pm \sqrt{b^2 - 4ac}}{2a}.$$

Therefore,

$$\lambda = \frac{5 \pm \sqrt{25 - 16}}{2}$$

$$= \frac{5 \pm \sqrt{9}}{2}.$$

Hence,

$$\lambda_1 = 4 \quad \text{and} \quad \lambda_2 = 1.$$

These roots of the characteristic equation are known as the characteristic roots, the eigenvalues, or the latent roots. If $[A]$ is symmetric, the roots will be real; when the matrix $[A]$ is also positive definite, as for the case in this simple example, the characteristic roots will be positive.

In a system of homogeneous linear equations there will be one solution vector, called a characteristic or latent vector, corresponding to each characteristic root. In the principal components analysis this vector contains the loadings for the particular component over all the variables.

Gould (1966b) has presented an excellent discussion of the geometrical interpretations of eigenvalues and has reviewed their application in geographic research.

As with all the topics in this appendix, we barely have exposed the surface of the mathematics of matrix algebra. The properties of orthogonal matrices and of matrices containing not only real but complex elements, for example, are important in certain lines of analysis, but they have been ignored here. This appendix has sought only to give an introduction to the different areas of mathematics, however, and hopefully the reader will pursue his own studies far beyond the limited boundaries of this cursory treatment.

APPENDIX B
GLOSSARY OF TERMS
AND SYMBOLS

Operators

$$\sum_{i=1}^{N} X_i \qquad \text{represents the } sum \text{ of } N \text{ values of } X.$$

$$\sum_{i=1}^{4} X_i = X_1 + X_2 + X_3 + X_4.$$

$$\sum_{i=1}^{N} k X_i = k \sum_{i=1}^{N} X_i, \qquad \text{where } k \text{ is a constant.}$$

$$\sum_{i=1}^{N} k = Nk, \qquad \text{where } k \text{ is a constant.}$$

$$\sum_{j=1}^{m} \sum_{i=1}^{N} X_{ij} = (X_{11} + X_{12} + \ldots + X_{1N}) + (X_{21} + X_{22} + \ldots X_{2N}) + \ldots$$
$$+ (X_{m1} + X_{m2} + \ldots + X_{mN}).$$

$$\prod_{i=1}^{N} X_i \qquad \text{represents the } product \text{ of } N \text{ values of } X.$$

$$\prod_{i=1}^{4} X_i = X_1 \cdot X_2 \cdot X_3 \cdot X_4.$$

$$\sum_{i=1}^{N} k X_i = k^N \left(\prod_{i=1}^{N} X_i \right), \qquad \text{where } k \text{ is a constant.}$$

$$\prod_{i=1}^{N} k = k^N, \qquad \text{where } k \text{ is a constant.}$$

257

Mathematical symbols

$a < b$	a less than b.
$a \leqslant b$	a less than or equal to b.
$a > b$	a greater than b.
$a \geqslant b$	a greater than or equal to b.
$a \neq b$	a not equal to b.
$a \approx b$	a is approximately equal to b.
$\infty, -\infty$	plus and minus infinity; increasing or decreasing without bounds.
$A \cup B$	the union of the events A and B.
$A \cap B$	the intersection of the events A and B.
$\log x$	logarithm of x to the base 10.
$\ln x$	logarithm of x to the base e; the natural logarithm of x.
dy/dx	derivative of y with respect to x.
dy	differential of y.
$\int$	integral sign.

Mathematical constants and functions

e Base of natural logarithms; $\lim_{l \to 0} (1 + l)^{1/l}$; value of 2.71828. ...

π Ratio of the circumference of a circle to its diameter; value of 3.14159. ...

radian   An angular measure relating the length of arc subtended on a circle by an angle at the center to the radius of the circle. 1 rad $= 360°/2\pi$.

$x!$ Read as x factorial; equals $x(x - 1)(x - 2)\ldots(2)(1)$. Thus $6! = 6(5)(4)(3)(2)(1) = 720$.

$\binom{n}{r} = C_r^n = \dfrac{r!(n - r)!}{n!}$ The combinatorial expression for the number of combinations of r objects which can be drawn from a total of N. Thus, $C_2^4 = 4!/(2!2!) = 6$.

Γ The gamma function: For any positive integer n, $\Gamma(n) = (n - 1)!$. For positive real values of p, $\Gamma(p) = \int_0^\infty x^{p-1} e^{-x} \, dx$. The function is tabled.

Number concepts

Integer   Any number in the infinite set of whole numbers 1, 2, 3, 4, 5, ...

Rational number   The quotient of two integers; may be written as a fraction or as a number with decimal places.

Irrational number   A number which cannot be expressed as a fraction or rational, for example, the values of e and π.

Real numbers   The infinite set of all positive and negative numbers, rational and irrational, including zero.

Complex numbers   Numbers of the form $\beta + i\alpha$, where β and α are real numbers and i is the imaginary operator such that $i = \sqrt{-1}$.

$|a|$ The absolute value of a; the numerical value of a regardless of its sign.

REFERENCES

Ahmad, Q. (1965), "Indian Cities: Characteristics and Correlates." *Research Paper 102*, Department of Geography, University of Chicago.

Aitchison, J., and Brown, J. A. C. (1957), *The Log-Normal Distribution*. Cambridge: Cambridge University Press.

Alexandersson, G. (1956), *The Industrial Structure of American Cities*. Lincoln, Nebraska: University of Nebraska Press.

Allen, G. R. (1954), "The 'Courbe des Populations,' A Further Analysis." *Bulletin of the Oxford University Institute of Statistics*, Vol. 16, pp. 179–189.

Allen, R. G. D. (1959), *Mathematical Economics*. London: Macmillan and Co. Ltd.

Alonso, W. (1964), "Location Theory," in Friedmann, J., and Alonso, W. (eds.) *Regional Development and Planning*. Cambridge, Mass.: The M. I. T. Press.

Anderson, T. W. (1958), *Introduction to Multivariate Statistical Analysis*. New York: John Wiley & Sons, Inc.

Anscombe, F. J. (1950), "Sampling Theory of the Negative Binomial and Logarithmic Series Distributions." *Biometrika*, Vol. 37, pp. 358–382.

Armstrong J. Scott (1967), "Derivation of Theory by Means of Factor Analysis or Tom Swift and His Electric Factor Analysis Machine." *The American Statistician*, Vol. 21, pp. 17–21.

Artle, R. (1965), *Studies in the Structure of the Stockholm Economy*. Ithaca: Cornell University Press.

Attneave, F. (1950), "Dimensions of Similarity." *American Journal of Psychology*, Vol. 63, pp. 516–556.

Bachi, R. (1958), "Statistical Analysis of Geographic Series." *Bulletin de l'Institut International de Statistique*, Vol. 36, pp. 229–240.

Bachi, R. (1963), "Standard Distance Measures and Related Methods for Spatial Analysis." *Papers*, The Regional Science Association, Vol. 10, pp. 83–132.

Bailey, N. T. J. (1964), *The Elements of Stochastic Processes with Applications to the Natural Sciences*. New York: John Wiley & Sons, Inc.

Barber, N. F. (1961), *Experimental Correlograms and Fourier Transforms*. New York: Pergamon Press.

Bartlett, M. S. (1963), "The Spectral Analysis of Point Processes." *Journal Royal Statistical Society*, B, Vol. 25, pp. 264–296.

Bartlett, M. S. (1964), "The Spectral Analysis of Two-Dimensional Point Processes." *Biometrika*, Vol. 51, pp. 299–311.

Barton, D. E., and David, F. N. (1956), "Tests for Randomness of Points on a Line." *Biometrika*, Vol. 43, pp. 104–112.

Barton, D. E., and David, F. N. (1962), "Randomization Bases for Multivariate Tests: I The Bivariate Case; Randomness of *N* Points in a Plane." *Bulletin de l'Institut International de Statistique*, Vol. 39, pp. 455–466.

Batschelet, E. (1965), *Statistical Methods for the Analysis of Problems in Animal Orientation and Certain Biological Rhythms*. Washington, D.C.: The American Instistute of Biological Sciences.

Bell, W. H., and Stevenson, D. W. (1964), "An Index of Economic Health for Ontario Counties and Districts." *Ontario Economic Review*, Vol. 2, 7 pp.

Bendat, J. S., and Piersol, A. G. (1966), *Measurement and Analysis of Random Data*. New York: John Wiley & Sons, Inc.

Berry, B. J. L. (1960a), "An Inductive Approach to the Regionalization of Economic Development." In *Essays on Geography and Economic Development*, Research Paper 62, Department of Geography, University of Chicago.

Berry, B. J. L. (1960b), "The Impact of Expanding Metropolitan Communities Upon the Central Place Hierarchy." *Annals*, Association of American Geographers, Vol. 50, pp. 112–116.

Berry, Brian J. L. (1961a), "City Size Distributions and Economic Development." *Economic Development and Cultural Change*, Vol. 9, pp. 573–588.

Berry, B. J. L. (1961b), "A Statistical Analysis." *Atlas of Economic Development*, Research Paper 68, Department of Geography, University of Chicago. Part VIII.

Berry, B. J. L. (1961c), "A Method for Deriving Multi-Factor Uniform Regions." *Przeglad Geograficzny*, Vol. 33, pp. 263–282.

Berry, B. J. L. (1962), "Sampling, Coding, and Storing Flood Plain Data." *Agricultural Handbook*, United States Department of Agriculture, No. 237.

Berry, B. J. L. (1963), "Commercial Structure and Commercial Blight." *Research Paper 85*, Department of Geography, University of Chicago.

Berry, B. J. L. (1964), "Approaches to Regional Analysis: A Synthesis." *Annals*, Association of American Geographers, Vol. 54, pp. 2–11.

Berry, B. J. L. (1965), "Identification of Declining Regions: An Empirical Study of the Dimensions of Rural Poverty." In R. S. Thoman and W. D. Wood (eds.) *Areas of Economic Stress in Canada*. Kingston: Queen's University.

Berry, B. J. L. (1966), "Essays on Commodity Flows and the Spatial Structure of the Indian Economy." *Research Paper 111*, Department of Geography, University of Chicago.

Berry, B. J. L. (1967a), "The Mathematics of Economic Regionalization." *Proceedings*, Brno Conference on Economic Regionalization. Brno: Czech Academy of Sciences.

Berry, B. J. L. (1967b), "Grouping and Regionalizing: An Approach to the Problem of Using Multivariate Analysis." In W. L. Garrison and D. F. Marble (eds.), *Quantitative Geography*, Northwestern University Studies in Geography, No. 13, pp. 219–251.

Berry, B. J. L., and Garrison, W. L. (1958), "The Functional Bases of the Central Place Hierarchy." *Economic Geography*, Vol. 34, pp. 145–154.

Berry, Brian J. L., and Garrison, William L. (1958), "Alternate Explanations of Urban Rank-Size Relationships." *Annals*, Association of American Geographers, Vol. 48, pp. 83–91.

Berry, B. J. L., and Barnum, H. G. (1962), "Aggregate Relations and Elemental Components of Central Place Systems." *Journal of Regional Science*, Vol. 4, pp. 35–68.

Berry, B. J. L., Simmons, J. W., and Tennant, R. J. (1963), "Urban Population Densities: Structure and Change." *Geographical Review*, Vol. 53, pp. 389–405.

Berry, B. J. L., and Ray, D. M. (1966), "Multivariate Socio-Economic Regionalization: A Pilot Study in Central Canada." In T. Rymes and S. Ostry (eds.), *Regional Statistical Studies*. Toronto: University of Toronto Press.

Berry, B. J. L., and Marble, D. F. (1967), *Spatial Analysis. A Reader in Statistical Geography*. Englewood Cliffs, N. J.: Prentice-Hall, Inc.

Bharucha-Reid, A. T. (1960), *Elements of the Theory of Markov Processes and Their Applications*. New York: McGraw-Hill Book Co.

Biometrics (1967), Vol. 13, pp. 261–405. Series of papers on covariance analysis.

Birch, J. W. (1960), "A Note on the Sample-Farm Survey and Its Use as a Basis for Generalized Mapping." *Economic Geography*, Vol. 36, pp. 254–259.

Blackith, R. E. (1958), "Nearest-Neighbor Distance Measurements for the Estimation of Animal Populations." *Ecology*, Vol. 39, 150–157.

Blackman, R. B., and Tukey, J. W. (1959), *The Measurement of Power Spectra from the Point of View of Communications Engineering*. New York: Dover Publications, Inc.

Blalock, H. M. (1963), "Correlated Independent Variables: The Problem of Multicollinearity." *Social Forces*, Vol. 42, pp. 233–237.

Blalock, H. M. (1964), *Causal Inferences in Nonexperimental Research*. Chapel Hill: The University of North Carolina Press.

Blaut, J. M. (1959), "Micro-Geographic Sampling: A Quantitative Approach to Regional Agricultural Geography." *Economic Geography*, Vol. 35, pp. 79–88.

Board, C. (1964), "A Sample Survey to Assess the Effect on Bantu Agriculture of the Rehabilitation Programme." *South African Journal of Science*, Vol. 60, pp. 229–235.

Bowden, L. W. (1965), "Diffusion of the Decision to Irrigate: Simulation of the Spread of a New Resource Management Practice in the Colorado Northern High Plains." *Research Paper 97*, Department of Geography, University of Chicago.

Boyce, D. E. (1965), "The Effect of Direction and Length of Person Trips on Urban Travel Patterns." *Journal of Regional Science*, Vol. 6, pp. 65–80.

Brown, G. W. (1947), "Discriminant Functions." *Annals of Mathematical Statistics*, Vol. 18, pp. 514–528.

Brown, L. (1963), "The Diffusion of Innovation: A Markov Chain-Type Approach." *Discussion Paper No. 3*, Department of Geography, Northwestern University.

Brown, L. (1965), "Models for Spatial Diffusion Research—A Review." *Technical Report No. 3*, Spatial Diffusion Study, Department of Geography, Northwestern University.

Brown, L. (1968), "Diffusion Dynamics: A Review and Revision of the Quantitative Theory of the Spatial Diffusion of Innovation." Forthcoming in *Lund Studies in Geography*.

Brown, S. E., and Trott, C. E. (1968), "Grouping Tendencies in an Economic Regionalization of Poland." *Annals*, Association of American Geographers, forthcoming.

Brush, J. E. (1953), "The Hierarchy of Central Places in Southwestern Wisconsin." *Geographical Review*, Vol. 43, pp. 380–402.

Bryan, J. G. (1951), "The Generalized Discriminant Function: Mathematical Foundation and Computational Routine." *Harvard Educational Review*, Vol. 21, pp. 90–95.

Bunge, W. (1962), "Theoretical Geography." *Lund Studies in Geography Series C.* Vol. 1, 210 pp.

Burton, I. (1963), The Quantitative Revolution and Theoretical Geography." *The Canadian Geographer*, Vol. 7, pp. 151–162.

Cain, S. A., and de Oliveira Castro, G. M. (1959), *Manual of Vegetation Analysis.* New York: Harper & Row, Publishers, Inc.

Campbell, N. (1952), *What is Science?* New York: Dover Publications, Inc.

Carey, G. W. (1966), "The Regional Interpretation of Manhattan Population and Housing Patterns Through Factor Analysis." *The Geographical Review*, Vol. 56, pp. 551–569.

Casetti, E. (1964a), "Multiple Discriminant Functions." *Technical Report No. 11*, Computer Applications in the Earth Sciences Project, Department of Geography, Northwestern University.

Casetti, E. (1964b), "Classificatory and Regional Analysis by Discriminant Iterations." *Technical Report No.12*, Computer Applications in the Earth Sciences Project, Department of Geography, Northwestern University.

Casetti, E. (1966), "Analysis of Spatial Association by Trigonometric Polynomials." *The Canadian Geographer*, Vol. 10, pp. 199–204.

Cattell, R. B. (1965), "Factor Analysis: An Introduction to Essentials I and II." *Biometrics*, Vol. 21, pp. 190–215, 405–435.

Cattell, R. B., ed. (1966), *Handbook of Multivariate Experimental Psychology.* Chicago: Rand McNally & Co.

Chacko, V. J., and Negi, G. S. (1965), "A Statistical Study of the Spatial Distribution of Dead Trees in a Casuarina Plantation." *Sankhya*, Series B, Vol. 27, pp. 211–224.

Chapman, M. (1963), "Sampling Methods for Geographers." *Pacific Viewpoint*, Vol. 4, pp. 155–163.

Chorley, R. J. (1964), "Geography and Analogue Theory." *Annals*, Association of American Geographers, Vol. 54, pp. 127–137.

Chorley, R. J., and Haggett, P. (1965), "Trend Surface Mapping in Geographical Research." *Publication No. 37*, Institute of British Geographers, pp. 47–67.

Chorley, R. J., and Haggett, P., eds. (1967), *Models in Geography*. London: Methuen and Co., Ltd.

Choynowski, M. (1959), "Maps Based on Probabilities." *Journal of American Statistical Association*, Vol.. 54, pp. 385–388.

Christ, C. F. (1966), *Econometric Models and Methods.* New York: John Wiley & Sons, Inc.

Christensen, W. I., and Bryson, R. A. (1966), "An Investigation of the Potential of Component Analysis for Weather Classification." *Monthly Weather Review*, Vol. 94, pp. 697–709.

Churchman C. W., and Ratoosh, P. (1959), *Measurement: Definitions and Theories.* New York: John Wiley & Sons, Inc.

Clark, P. J. (1956), "Grouping in Spatial Distributions." *Science*, Vol. 123, pp. 373–374.

Clark, P. J. and Evens, F. C. (1954), "Distance to Nearest Neighbor as a Measure of Spatial Relationships in Populations." *Ecology*, Vol. 35, pp. 445–453.

Clark, P. J., and Evans, F. C. (1955), "On Some Aspects of Spatial Patterns in Biological Populations." *Science*, Vol. 121, pp. 397–398.

Cochran, W. G. (1953), *Sampling Techniques*. New York: John Wiley & Sons, Inc.

Cochran, W. G., Mosteller, F., and Tukey, J. W. (1954), "Principles of Sampling." *Journal of the American Statistical Association*, Vol. 49, pp. 13–35.

Cole, J. P. (1966), "Set Theory and Geography." *Bulletin of Quantitative Data for Geographers*, No. 2, Department of Geography, Nottingham University.

Coleman, J. S. (1964), *Introduction to Mathematical Sociology*. New York: The Free Press.

Cooley, W. W., and Lohnes, P. R. (1962), *Multivariate Procedures for the Behavioral Sciences*. New York: John Wiley & Sons, Inc.

Court, A. (1964), "The Elusive Point of Minimum Travel." *Annals*, Association of American Geographers, Vol. 54, pp. 400–403.

Court, A. (1966), "Population Distributions and Self-Potential." Mimeographed paper, Department of Geography San Fernando Valley State College.

Cox, D. R., and Lewis, A. W. (1966), *The Statistical Analysis of Series of Events*. London: Methuen and Co., Ltd.

Cox, K. R. (1968), "Suburbia and Voting Behavior in the London Metropolitan Area." *Annals*, Association of American Geographers, Vol. 58.

Cunnyngham, J. (1963), "The Spectral Analysis of Economic Time Series." *Working Paper No. 14*, Bureau of the Census, U.S. Department of Commerce.

Curry, L. (1962), "The Geography of Service Centers Within Towns: The Elements of an Operational Approach." *Lund Studies in Geography, Series B*, Vol. 24, pp. 31–53.

Curry, L. (1963), "Regional Variation in the Seasonal Programming of Livestock Farms in New Zealand." *Economic Geography*, Vol. 39, pp. 96–118.

Curry, L. (1964), "The Random Spatial Economy: An Exploration in Settlement Theory." *Annals*, Association of American Geographers, Vol. 54, pp. 138–146.

Curry, L. (1966a), "Seasonal Programming and Bayesian Assessment of Atmospheric Resources." In W. R. D. Sewell (ed.), *Human Dimensions of Weather Modification. Research Paper 105*, Department of Geography, University of Chicago, pp. 127–138.

Curry, L. (1966b), "A Note on Spatial Association." *The Professional Geographer*, Vol. 18, pp. 97–99.

Curry, L. (1966c), "Chance and Landscape." In J. W. House (ed.), *Northern Geographical Essays in Honour of G. H. J. Daysh*, Newcastle-upon-Tyne: Oriel Press, pp. 40–55.

Curry, L. (1966d), "Central Places in the Random Spatial Economy." Paper presented at National Science Foundation sponsored Seminar on Quantitative Methods in Geography, Columbus, Ohio. Forthcoming in *Journal of Regional Science*.

Curry, L. (1967), "Quantitative Geography." *The Canadian Geographer*, Vol. 11, pp. 265–279.

Curtis, J. T., and McIntosh, R. P. (1950), "The Interrelations of Certain Analytic and Synthetic Phytosociological Characters." *Ecology*, Vol. 31, pp. 434–455.

Dacey, M. F. (1960a), "A Note on the Derivation of Nearest-Neighbor Distances." *Journal of Regional Science*, Vol. 2, pp. 81–87.

Dacey, M. F. (1960b), "The Spacing of River Towns." *Annals*, Association of American Geographers, Vol. 50, pp. 59–61.

Dacey, M. F. (1962), "Analysis of Central Place and Point Patterns by a Nearest-Neighbor Method." *Lund Studies in Geography, Series B*, Vol. 24, pp. 55–75.

Dacey, M. F. (1963), "Order Neighbor Statistics for a Class of Random Patterns in Multidimensional Space." *Annals*, Association of American Geographers, Vol. 53, pp. 505–515.

Dacey, M. F. (1964a), "Two-Dimensional Random Point Patterns: A Review and an Interpretation." *Papers*, The Regional Science Association, Vol. 13, pp. 41–55.

Dacey, M. F. (1964b), "Modified Poisson Probability Law for Point Pattern More Regular Than Random." *Annals*, Association of American Geographers, Vol. 54, pp. 559–565.

Dacey, M. F. (1965a), "Order Distance in an Inhomogeneous Random Point Pattern." *The Canadian Geographer*, Vol. 9, pp. 144–153.

Dacey, M. F. (1965b), "A Review on Measures of Contiguity for Two and k-Color Maps." *Technical Report No. 2*, Spatial Diffusion Study, Department of Geography, Northwestern University.

Dacey, M. F. (1965c), "Measures of Distance From a Randomly Located Point to Neighboring Lattice Points for Rectangular and Hexagonal Point Lattices." *Technical Report No. 3*, Geographical Information Systems Project, Department of Geography, Northwestern University.

Dacey, M. F. (1965d), "Numerical Measures of Random Sets." *Technical Report No. 5*, Geographical Information Systems Project, Department of Geography, Northwestern University.

Dacey, M. F. (1965e), "Four Properties of Two-Dimensional Random Point Patterns." *Technical Report No. 6*, Geographical Information Systems Project, Department of Geography, Northwestern University.

Dacey, M. F. (1965f), "A Stochastic Model of Economic Regions." *Discussion Paper No. 4*, Department of Geography, Northwestern University.

Dacey, M. F. (1966a), "A Compound Probability Law for a Pattern More Dispersed than Random and with Areal Inhomogeneity. *Economic Geography*, Vol. 42, pp. 172–179.

Dacey, M. F. (1966b), "A County-Seat Model for the Areal Pattern of an Urban System." *Geographical Review*, Vol. 56, pp. 527–542.

Dacey, M. F. (1966c), "A Probability Model for Central Place Locations." *Annals*, Association of American Geographers, Vol. 56, pp. 550–568.

Dacey, M. F. (1966d), "A Model for the Areal Pattern of Retail and Service Establishments Within an Urban Area." *Technical Report No. 2*, Urban and Transportation Information Systems Project, Department of Geography, Northwestern University.

Dacey, M. F. (1967), "An Empirical Study of the Areal Distribution of Houses in Puerto Rico." Mimeographed paper, Northwestern University.

Dacey, M. F., and Tung, T. (1962), "The Identification of Randomness in Point Patterns." *Journal of Regional Science*, Vol. 4, pp. 83–96.

Dalenius, T., Hájek, J., and Zubrzycki, S. (1961), "On Plane Sampling and Related

Geometrical Problems." *Proc. Fourth Berkeley Symposium on Mathematical Statistics and Probability*, Vol. 4, pp. 125–150.

Decker, W. L. (1952), "Hail-Damage Frequencies for Iowa and a Method of Evaluating the Probability of a Specified Amount of Hail Damage." *Trans. American Geophysical Union*, Vol. 33, pp. 204–210.

Downs, R. M. (1967), "Approaches to, and Problems in the Measurement of Geographic Space Perception." *Seminar Series Paper No. 9*, Department of Geography, University of Bristol.

Draper, N. R., and Smith, H. (1966), *Applied Regression Analysis*. New York: John Wiley & Sons, Inc.

Duncan, O. D., et al. (1961), *Statistical Geography: Problems in Analyzing Areal Data*. New York: The Free Press.

Dwyer, P. S. (1945), "The Square Root Method and its Use in Correlation and Regression." *Journal of the American Statistical Association*, Vol. 40, pp. 493–503.

Dwyer, P. S. (1951), *Linear Computations*. New York: John Wiley & Sons, Inc.

El-Kammash, M. M. (1963), "On the Measurement of Economic Development Using Scalogram Analysis." *Papers*, The Regional Science Association, Vol. 11, pp. 309–334.

Ezekiel, M., and Fox, K. A. (1959), *Methods of Correlation and Regression Analysis*, Third edition. New York: John Wiley & Sons, Inc.

Faddeeva, V. N. (1959), *Computational Methods of Linear Algebra*. New York: Dover Publications, Inc.

Fairbairn, K. J., and Robinson, G. (1967a), "Towns and Trend-Surfaces in Gippsland, Australia." Mimeographed paper, Department of Geography, Monash University, Melbourne.

Fairbairn, K. J., and Robinson, G. (1967b), "An Application of Trend-Surface Mapping to the Distribution of Residuals from a Regression." Mimeographed paper, Department of Geography, Monash University, Melbourne.

Feller, W. (1950), *An Introduction to Probability Theory and Its Applications*. New York: John Wiley & Sons, Inc.

Firher, J. C. (1966), *Yugoslavia—A Multinational State*. San Francisco: Chandler Publishing Co.

Fisher, R. A. (1936), "The Use of Multiple Measurements in Taxonomic Problems." *Ann. Eugenics*, Vol. 7, pp. 179–188.

Fraser, D. A. S. (1957), *Nonparametric Methods in Statistics*. New York: John Wiley & Sons, Inc.

Fraser, D. A. S. (1958), *Statistics: An Introduction*. New York: John Wiley & Sons, Inc.

Frisch, R. (1966), *Maxima and Minima. Theory and Economic Applications*. Chicago: Rand McNally & Co.

Fruchter, B. (1954), *Introduction to Factor Analysis*. Princeton, N.J.: D. Van Nostrand Co., Inc.

Fuchs, R. J. (1960), "Intraurban Variation of Residential Quality." *Economic Geography*, Vol. 36, pp. 313–325.

Galliher, H. P. (1959), "Simulation of Random Processes." In *Notes on Operations Research*. Cambridge, Mass.: The M.I.T. Press, pp. 231–246.

Garrison, W. L. (1956a), "Some Confusing Aspects of Common Measurements." *The Professional Geographer*, Vol. 8, pp. 4–5.

Garrison, W. L. (1956b), *The Benefits of Rural Roads to Rural Property*. Seattle: Washington State Council for Highway Research.

Garrison, W. L. (1962), "Toward a Simulation Model of Urban Growth and Development." *Lund Studies in Geography, Series B*, Vol. 24, pp. 91–108.

Garrison, W. L., and Marble, D. F. (1964), "Factor-Analytic Study of the Connectivity of a Transport Network." *Papers*, Regional Science Association, Vol. 12, pp. 231–238.

Garrison, W. L., and Marble, D. F. (eds.) (1967), *Quantitative Geography*, Northwestern University Studies in Geography, No. 13.

Gauthier, H. L. (1966), "Transportation and the Growth of the São Paulo Economy." Paper presented at Regional Science meeting, St. Louis. Forthcoming in *Journal of Regional Science*.

Geary, R. C. (1954), "The Contiguity Ratio and Statistical Mapping." *The Incorporated Statistician*, Vol. 5, pp. 115–141.

Geary, R. C. (1963), "Some Remarks About Relations Between Stochastic Variables." *Revue de l'Institut International de Statistique*, Vol. 31, pp. 163–181.

Geisler, M. A, Haythorn, W. W., and Steger, W. A. (1962), "Simulations and Logistics System Laboratory." *Rand Corporation Memorandum*, RM-3281.

Getis, A. (1964), "Temporal Land-Use Patterns Analysis with the Use of Nearest-Neighbor and Quadrat Methods." *Annals*, Association of American Geographers, Vol. 54, pp. 391–399.

Gibrat, R. (1931), *Les inégalités économiques*. Paris: Sirey.

Gittus, E. (1964), "The Structure of Urban Areas: A New Approach." *The Town Planning Review*, Vol. 35, pp. 5–20.

Glasser, G. J. (1962), "Variance Formulas for the Mean Difference and Coefficient of Concentration." *Journal of the American Statistical Association*, Vol. 57, pp. 648–654.

Goldberg, S. (1960), *Probability: An Introduction*. Englewood Cliffs, N.J.: Prentice-Hall, Inc.

Golledge, R. G., and Amadeo, D. M. (1966), "Some Introductory Notes on Regional Division and Set Theory." *The Professional Geographer*, Vol. 18, pp. 14–19.

Golledge, R. G. (1967), "A Conceptual Framework of a Market Decision Process." Forthcoming in *Journal of Regional Science*.

Goodall, D. W. (1952), "Some Considerations in the Use of Point Quadrats for the Analysis of Vegetation." *Australian Science Journal*, Series B, Vol. 5, pp. 1–61.

Goodall, D. W. (1954), "Objective Methods for the Classification of Vegetation, III. An Essay in the Use of Factor Analysis." *Australian Journal of Botany*, Vol. 2, pp. 304–324.

Gould, P. R. (1963), "Man Against his Environment: A Game Theoretic Framework." *Annals*, Association of American Geographers, Vol. 53, pp. 290–97.

Gould, P. R. (1966a), "Space Searching Procedures in Geography and the Social Sciences." *Working Papers No. 1*, Science Research Institute, University of Hawaii.

Gould, P. R. (1966b), "On the Geographic Interpretation of Eigenvalues." Mimeographed paper, Department of Geography, Pennsylvania State University. Forthcoming in *Publications of the Institute of British Geographers.*

Gould, P. R. (1966c), "On Mental Maps." *Discussion Paper No. 9*, Michigan Inter-University Community of Mathematical Geographers.

Granger, C. W. (1964), *Spectral Analysis of Economic Time Series.* Princeton, N. J.: Princeton University Press.

Gregory, S. (1963), *Statistical Methods and the Geographer.* London: Longmans, Green & Co., Ltd.

Greig-Smith, P. (1964), *Quantitative Plant Ecology*, Second edition. London: Butterworth & Co., Ltd.

Gurevich, B. L., and Saushkin, Y. G. (1966), "The Mathematical Method in Geography." *Soviet Geography*, Vol. 7, pp. 3–35.

Gurevich, B. L. (1967), "The Density of Population of a City and the Density of Probability of a Random Magnitude." *Soviet Geography*, Vol. 8, pp. 722–730.

Guttman, L. (1967), "The Development of Nonmetric Space Analysis: A Letter to Professor John Ross." *Multivariate Behavioral Research*, Vol. 2, pp. 71–82.

Hadden, J. K., and Borgatta, E. F. (1965), *American Cities: Their Social Characteristics.* Chicago: Rand McNally & Co.

Hägerstrand, T. (1952), "The Propagation of Innovation Waves." *Lund Studies in Geography, Series B*, Vol. 4, pp. 3–19.

Hägerstrand, T. (1953), *Innovationsförloppet ur korologisk synpunkt.* Lund: Gleerups Förlag.

Hägerstrand, T. (1957), "Migration and Area: Survey of a Sample of Swedish Migration Fields and Hypothetical Considerations on their Genesis."*Lund Studies in Geography, Series B*, Vol. 13, pp. 27–158.

Hägerstrand, T. (1965), "A Monte Carlo Approach to Diffusion." *Achives Europeenes de Sociologie*, Vol. 6, pp. 43–67.

Hägerstrand, T. (1966), "Aspects of the Spatial Structure of Social Communication and the Diffusion of Information." *Papers*, Regional Science Association, Vol. 16, pp. 27–42.

Hägerstrand, T. (1967), "On Monte Carlo Similation of Diffusion." In W. L. Garrison and D. F. Marble (eds.), *Quantitative Geography*, Northwestern University Studies in Geography, No. 13, pp. 1–32.

Haggett, P. (1963), "Regional and Local Components in Land-Use Sampling." *Erdkunde*, Vol. 17, pp. 108–114.

Haggett, P. (1964), "Regional and Local Components in the Distribution of Forested Areas in South East Brazil: A Multivariate Approach." *Geographical Journal*, Vol. 130, pp. 365–380.

Haggett, P. (1965), *Locational Analysis in Human Geography.* New York: St. Martin's Press, Inc.

Haggett, P. (1967), "Trend Surface Mapping in Interregional Comparisons." Paper presented at The Regional Science Association Meeting, The Hague.

Haggett, P., and Board, C. (1964), "Rotational and Parallel Traverses in the Rapid Integration of Geographic Areas." *Annals*, Association of American Geographers, Vol. 54, pp. 406–410.

Haggett, P., and Chorley, R. J. (1967), "Models, Paradigms and the New Geography." in R. J. Chorley and P. Haggett (eds.) *Models in Geography.* London: Methuen and Co., Ltd., pp. 19–41.

Hagood, M. J. (1943), "Statistical Methods for Delineation of Regions Applied to Data on Agriculture and Population." *Social Forces*, Vol. 21, pp. 287–297.

Hald, A. (1952), *Statistical Theory with Engineering Applications.* New York: John Wiley & Sons, Inc.

Hammersley, J. M., and Handscomb, D. C. (1964), *Monte Carlo Methods.* London: Methuen and Co., Ltd.

Hannan, E. J. (1960), *Time Series Analysis.* London: Methuen and Co., Ltd.

Hansen, W. B. (1961), "An Approach to the Analysis of Metropolitan Residential Extension." *Journal of Regional Science*, Vol. 3, pp. 37–55.

Harbaugh, J. W., and Preston, F. W. (1965), "Fourier Series Analysis in Geology." *Short Course and Symposium on Computers and Computer Applications in Mining and Exploration*, Vol. 1. University of Arizona.

Hare, F. K. (1958), "The Quantitative Representation of the North Polar Pressure Fields." In *Polar Atmosphere Symposium, Part I, Meteorology Section*, New York: Pergamon Press.

Harman, H. H. (1960), *Modern Factor Analysis*, Chicago: The University of Chicago Press.

Harris, B. (1964), "A Note on the Probability of Interaction at a Distance." *Journal of Regional Science*, Vol. 5, pp. 31–35.

Harris, B. (1966), "The Uses of Theory in the Simulation of Urban Phenomena." *Highway Research Record*, No. 126, pp. 1–16.

Harris, C. W., ed. (1963), *Problems in Measuring Change.* Madison: The University of Wisconsin Press.

Hart, J. F. (1954), "Central Tendency in Areal Distributions." *Economic Geography*, Vol. 30, pp. 48–59.

Hartman, G. W., and Hook, J. C. (1956), "Substandard Housing in the United States: A Quantitative Analysis." *Economic Geography*, Vol. 32, pp. 95–114.

Harvey, D. W. (1966), "Geographical Processes and the Analysis of Point Patterns." *Transactions and Papers*, Institute of British Geographers, No. 40, pp. 81–95.

Harvey, D. W. (1967), "Some Problems in the Use of the Neyman Type A and Negative Binomial Distribution." Unpublished paper, University of Bristol.

Harvey, D. (1968), "Behavioural Postulates and the Construction of Theory in Human Geography." Paper forthcoming in Proceedings of the Third Anglo-Polish Geography Seminar.

Henshall, J. D. (1966), "The Demographic Factor in the Structure of Agriculture in Barbados." *Publication No. 38*, Institute of British Geographers, pp. 183–195.

Henshall, J. D., and King, L. J. (1966), "Some Structural Characteristics of Peasant Agriculture in Barbados." *Economic Geography*, Vol. 42, pp. 74–84.

Hidore, J. J. (1963), "The Relations Between Cash-Grain Farming and Landfarms." *Economic Geography.* Vol. 39, pp. 84–89.

Hirshleifer, J. (1961), "The Bayesian Approach to Statistical Decision: An Exposition." *Journal of Business*, Vol. 34, pp. 471–489.

Hodge, G. (1963), "Use and Mis-use of Measurement Scales in City Planning." *Journal of American Institute of Planners*, Vol. 29, pp. 112–121.

Hodge, G. (1965), "The Prediction of Trade Center Viability in the Great Plains." *Papers*, The Regional Science Association, Vol. 15, pp. 87–115.

Hodge, G. (1966), *The Identification of 'Growth Poles' in Eastern Ontario*. A Report to the Ontario Department of Economics and Development, Toronto.

Hodges, J. L., et al. (1950–1955), *Discriminatory Analysis Reports 1–10*. Project No. 21–49–004. School of Aviation Medicine, U.S.A.F. Randolph Field, Texas.

Hodges, J. L., and Lehmann, E. L. (1965), *Elements of Finite Probability*. San Francisco: Holden-Day, Inc.

Hoel, P. G. (1960), *Elementary Statistics*. New York: John Wiley & Sons, Inc.

Hogg, R. V., and Craig, A. T. (1959), *Introduction to Mathematical Statistics*. New York: The Macmillan Company.

Hohn, F. E. (1964), *Elementary Matrix Algebra*, Second edition. New York: The Macmillan Co.

Horn, L. H., and Bryson, R. A. (1960), "Harmonic Analysis of the Annual March of Precipitation." *Annals*, Association of American Geographers, Vol. 50, pp. 157–171.

Horst, P. (1965), *Factor Analysis of Data Matrices*. New York: Holt, Rinehart and Winston, Inc.

Hotelling, H. (1931), "The Generalization of Student's Ratio." *Annals Mathematical Statistics*, Vol. 2, pp. 360–378.

Hotelling, H. (1933), "Analysis of a Complex of Statistical Variables into Principal Components." *Journal of Educational Psychology*, Vol. 24, pp. 417–441, 498–520.

Hotelling, H. (1936), "Relations Between Two Sets of Variates." *Biometrika*, Vol. 28, pp. 321–377.

Hudson, J. C. (1966), "Maps and Spatial Processes Describable with Markov Chains." Mimeographed paper, Department of Geography, University of North Dakota.

Huntsberger, D. V. (1961), *Elements of Statistical Inference*. Boston: Allyn and Bacon, Inc.

Imbrie, J. (1963), "Factor and Vector Analysis Programs for Analyzing Geologic Data." *Technical Report No. 6*, Computer Applications in the Earth Sciences Project, Department of Geography, Northwestern University.

Imbrie, J. (1964), "Factor Analytic Model in Palaeoecology." In *Approaches to Palaeoecology*. New York: John Wiley & Sons, Inc., pp. 407–422.

Imbrie, J., and van Andee, T. H. (1964), "Vector Analysis of Heavy Mineral Data," *Bulletin*, Geological Society of America, Vol. 75. pp. 1131–1156.

Isard, W. (1960), *Methods of Regional Analysis: An Introduction to Regional Science*. New York: John Wiley & Sons, Inc.

Jenkins, G. M. (1961), "General Considerations in the Analysis of Spectra." *Technometrics*, Vol. 3, pp. 133–166. See also the other papers on spectral analysis in this volume.

Johnson, N. L., and Leone, F. C. (1964), *Statistics and Experimental Design*. Vol. I. New York: John Wiley & Sons, Inc.

Johnston, J. (1963), *Econometric Methods*. New York: McGraw-Hill Book Co.

Journal of Applied Statistics. (1965), Vol. 14. Special issue devoted to the theory of spectral analysis and its applications.

Kansky, K. (1963), "Structure of Transportation Networks. "*Research Paper No. 84*, Department of Geography, University of Chicago.

Kapteyn, J. C. (1903), *Skew Frequency Curves in Biology and Statistics.* Groningen: E. P. Noordhoff.

Kariel, H. G. (1963), "Selected Factors Areally Associated with Population Growth Due to Net Migration." *Annals*, Association of American Geographers, Vol. 53, pp. 210–223.

Kates, R. W. (1962), "Hazard and Choice Perception in Flood Plain Management." *Research Paper 78*, Department of Geography, University of Chicago.

Kates, R. W., and Wohlwill, J. F. (1966), "Man's Response to the Physical Environment." Special issue of *The Journal of Social Issues*, Vol. 22.

Kemeny, J. G. (1959), *A Philosopher Looks at Science.* Princeton, N. J.: D. Van Nostrand Co., Inc.

Kemeny, J. G., Snell, J. L., and Thompson, G. L. (1966), *Introduction to Finite Mathematics*, Second edition. Englewood Cliffs, N. J.: Prentice-Hall, Inc.

Kemp, C. D., and Kemp, A. W. (1956), "The Analysis of Point Quadrat Data." *Australian Journal of Botany*, Vol. 4, pp. 167–174.

Kendall, M. G. (1939), "The Geographical Distribution of Crop Productivity in England." *Journal of Royal Statistical Society*, A, Vol. 102, pp. 21–62.

Kendall, M. G. (1957), *A Course in Multivariate Analysis.* London: Charles Griffin & Co., Ltd.

Kendall, M. G., and Moran, P. A. P. (1963), *Geometrical Probability.* New York: Hafner Publishing Co.

King, L. J. (1961), "A Multivariate Analysis of the Spacing of Urban Settlements in the United States. "*Annals*, Association of American Geographers, Vol. 51, pp. 222–233.

King, L. J. (1962a), "A Quantitative Expression of the Pattern of Urban Settlements in Selected Areas of the United States." *Tijdschrift voor Economische en Sociale Geografie*, Vol. 53, pp. 1–7.

King, L. J. (1962b), "The Functional Role of Small Towns in Canterbury." *Proceedings, Third New Zealand Geography Conference*, pp. 139–149.

King, L. J. (1966), "Cross-Sectional Analysis of Canadian Urban Dimensions: 1951 and 1961." *Canadian Geographer*, Vol. 10, pp. 205–224.

King, L. J. (1967), "Discriminatory Analysis of Urban Growth Patterns in Ontario and Quebec, 1951–1961." *Annals*, Association of American Geographers, Vol. 57, pp. 566–578.

Kish, L. (1965), *Survey Sampling.* New York: John Wiley & Sons, Inc.

Knos, D. S. (1962), *Distribution of Land Values in Topeka, Kansas.* Lawrence: University of Kansas.

Krumbein, W. C. (1962), "Open and Closed Number Systems in Stratigraphic Mapping." *Bulletin American Assoc. Petrol. Geologists*, Vol. 46, pp. 2229–2245.

Krumbein, W. C. (1966), "A Comparison of Polynomial and Fourier Models in Map Analysis." *Technical Report No. 2*, ONR Task No. 388–078, Nonr-1228 (36), Department of Geology, Northwestern University.

Krumbein, W. C., and Graybill, F. A. (1965), *An Introduction to Statistical Models in Geology*. New York: McGraw-Hill Book Co.

Kruskal, J. B. (1964), "Multidimensional Scaling by Optimizing Goodness of Fit to a Nonmetric Hypothesis." *Psychometrika*, Vol. 29, pp. 1–27.

Kulldorff, G. (1955), "Migration Probabilities." *Lund Studies in Geography, Series B*, Vol. 14, 45 pp.

Kulldorff, G. (1961), *Contributions to the Theory of Estimation from Grouped and Partially Grouped Samples*. New York: John Wiley & Sons, Inc.

Lahey, J. G., et al. (1958), *Atlas of 500 mb Wind Characteristics for the Northern Hemisphere*. Madison: The University of Wisconsin Press.

Lakshmanan, T. T. (1964), "An Approach to the Analysis of Intraurban Location Applied to the Baltimore Region." *Economic Geography*, Vol. 40, pp. 348–370,

Lanczos, C. (1956), *Applied Analysis*. Englewood Cliffs, N.J.: Prentice-Hall, Inc.

Lawley, D. N., and Maxwell, A. E. (1963), *Factor Analysis as a Statistical Method*. London: Butterworth and Co. (Publishers) Ltd.

Lee, Y. W. (1960), *Statistical Theory of Communication*. New York: John Wiley & Sons, Inc.

Leese, J. A., and Epstein, E. S. (1963), "Application of Two-Dimensional Spectral Analysis to the Quantification of Satellite Cloud Photographs." *Journal of Applied Meteorology*, Vol. 2, pp. 629–644.

Lösch, A. (1954), *The Economics of Location*. New Haven: Yale University Press.

Lubischew, A. A. (1962), "On the Use of Discriminant Functions in Taxonomy." *Biometrics*, Vol. 18, pp. 455–477.

Lukermann, F., and Porter, P. W. (1960), "Gravity and Potential Models in Economic Geography." *Annals*, Association of American Geographers, Vol. 50, pp. 493–504.

Luneburg, R. K. (1947), *Mathematical Analysis of Binocular Vision*. Princeton, N.J. Princeton University Press.

Mackay, J. R. (1958), "Chi-Square As a Tool for Regional Studies." *Annals*, Association of American Geographers, Vol. 48, p. 164.

Mackay, J. R., and Berry, B. J. L. (1959), "Comments on Use of Chi-Square." *Annals*, Association of American Geographers, Vol. 49, p. 89.

MacRae, D. (1960), "Direct Factor Analysis of Sociometric Data." *Sociometry*, Vol. 23, pp. 360–371.

McCarty, H. H., Hook, J. C., and Knos, D. S. (1956), *The Measurement of Association in Industrial Geography*. Iowa City: Department of Geography, University of Iowa.

McConnell, H. (1966a), "Quadrat Methods in Map Analysis." *Discussion Paper No. 3*, Department of Geography, University of Iowa.

McConnell, H. (1966b), "A Statistical Analysis of Spatial Variability of Mean Topographic Slope on Stream-Dissected Glacial Materials." *Annals*, Association of American Geographers, Vol. 56, pp. 712–728.

McConnell, J. E. (1967), "The Middle East: Competitive or Complementary?" *Tijdschrift voor Economische en Sociale Geografie*, Vol. 58, pp. 82–93.

Mabogunje, A. L. (1965), "Urbanization in Nigeria: A Constraint on Economic Development." *Economic Development and Cultural Change*, Vol. 13, pp. 413–438.

Mahalanobis, P. C. (1936), "On the Generalized Distance in Statistics." *Proceedings Nat. Inst. Science India*, Vol. 12, pp. 49–55.

Mahalanobis, P. C. (1949), "Historical Notes on the D^2 Statistic." *Sankhya*, Vol. 9, pp. 237–240.

Malinvaud, E. (1966), *Statistical Methods in Econometrics*. Chicago: Rand McNally & Co.

Marble, D. F. (1964), "Two Computer Programs for the Analysis of Simple Markov Chains." *Discussion Paper No. 6*, Regional Science Research Institute.

Marble, D. F. (1967), "A Theoretical Exploration of Individual Travel Behavior." In W. L. Garrison and D. F. Marble (eds.), *Quantitative Geography*, Northwestern University Studies in Geography, No. 13, pp. 33–53.

Marschak, J. (1966), "Foreward: A Remark on Econometric Tools." In C. F. Christ, *Econometric Models and Methods*, New York: John Wiley & Sons, Inc.

Massey, F. J. (1951), "The Kolmogorov-Smirnov Test for Goodness of Fit." *Journal of the American Statistical Association*, Vol. 46, pp. 68–78.

Massey, W. F. (1965), "Principal Components Regression in Exploratory Statistical Research." *Journal of the American Statistical Association*, Vol. 60, pp. 234–256.

Matern, B. (1960), "Spatial Variation." *Meddelanden Fran Statens Skogsforsknings-institut*, Vol. 49, pp. 1–144.

Matui, I. (1932), "Statistical Study of the Distribution of Scattered Villages in Two Regions of the Tonami Plain, Toyama Prefecture." *Japanese Journal of Geology and Geography*, Vol. 9, pp. 251–256.

Maxwell, A. E. (1961), *Analyzing Qualitative Data*. London: Methuen and Co., Ltd.

Mayfield, R. C. (1963), "The Range of a Central Good in the Indian Punjab." *Annals*, Association of American Geographers, Vol. 53, pp. 39–49.

Mayfield, R. C. (1967), "A Central-Place Hierarchy in Northern India." In W. L. Garrison and D. F. Marble (eds), *Quantitative Geography*, Northwestern University Studies in Geography, No. 13, pp. 120–166.

Medvedkov, Y. V. (1964), "Applications of Mathematics to Some Problems in Economic Geography." *Soviet Geography*, Vol. 5, pp. 36–53.

Medvedkov, Y. V. (1967), "Concept of Entropy in Settlement Pattern Analysis." *Papers*, The Regional Science Association, Vol. 18, pp. 165–168.

Megee, M. (1963), "Social and Economic Factors in the Differential Growth of Mexican States." *Papers*, First Latin American Conference on Regional Science, Caracas, pp. 143–178.

Megee, M. (1964), "Factor Analysis in Hypothesis Testing and Decision-Making." *The Professional Geographer*, Vol. 16, pp. 24–27.

Megee, M. (1965a), "On Economic Growth and the Factor Analysis Method." *Southern Economic Journal*, Vol. 31, pp. 215–228.

Megee, M. (1965b), "Economic Factors and Economic Regionalization in the United States." *Geografiska Annaler*, Vol. 47B, pp. 125–137.

Melton, M. A. (1957), "An Analysis of the Relations Among Elements of Climate, Surface Properties, and Geomorphology." *Technical Report No. 11*, ONR Project NR 389–042, Department of Geology, Columbia University.

Merriam, D. F., ed. (1966), "Computer Applications in the Earth Sciences: Colloquium on Classification Procedures." *Computer Contribution 7*, State Geological Survey, University of Kansas.

Miller, R. G. (1962), "Statistical Prediction by Discriminant Analysis." *Meteorological Monographs*, American Meteorological Society, Vol. 4.

Miller, R. L., and Kahn, J. S. (1962), *Statistical Analysis in the Geological Sciences*. New York: John Wiley & Sons, Inc.

Minnick, R. F. (1964), "A Method for the Measurement of Areal Correspondence." *Papers Michigan Academy of Science, Arts, and Letters*, Vol. 49, pp. 333–344.

Mises, R. von. (1945), "On the Classification of Observation Data into Distinct Groups." *Annals Mathematical Statistics*, Vol. 16, pp. 68–73.

Mode, E. B. (1966), *Elements of Probability and Statistics*, Englewood Cliffs, N.J.: Prentice-Hall, Inc.

Moore, P. G. (1954), "Spacing in Plant Populations." *Ecology*, Vol. 35, pp. 222–227.

Moran, P. A. P. (1948), "The Interpretation of Statistical Maps." *Journal of the Royal Statistical Society*, Series B, Vol. 10, pp. 245–251.

Morisita, M. (1957), "A New Method for the Estimation of Density by the Spacing Method Applicable to Nonrandomly Distributed Populations." *Seiro-Seitai*, Vol. 7, pp. 134–144.

Morisita, M. (1959), "Measuring the Dispersion of Individuals and Analysis of the Distributional Pattern." *Memoirs of the Faculty of Science of Kyushu University*, Series E, Vol. 2, pp. 215–235.

Morrill, R. L. (1962), "Simulation of Central Place Patterns over Time." *Lund Studies in Geography*, Series B, Vol. 24, pp. 109–120.

Morrill, R. L. (1963), "The Development of Spatial Distributions of Towns in Sweden: An Historical-Predictive Approach." *Annals*, Association of American Geographers, Vol. 53, pp. 1–14.

Morrill, R. L. (1965a), "Migration and the Spread and Growth of Urban Settlement." *Lund Studies in Geography*, Series B, Vol. 26.

Morrill, R. L. (1965b), "Expansion of the Urban Fringe: A Simulation Experiment." *Papers*, Regional Science Association, Vol. 15, pp. 185–199.

Morrill, R. L., and Pitts, F. R. (1967), "Marriage, Migration, and the Mean Information Field: A Study in Uniqueness and Generality." *Annals*, Association of American Geographers, Vol. 57, pp. 401–422.

Morrison, D. F. (1967), *Multivariate Statistical Methods*. New York: McGraw-Hill Book Co.

Morrissett, I. (1958), "The Economic Structure of American Cities." *Papers*, The Regional Science Association, Vol. 4, pp. 239–258.

Moser, C. A., and Scott, W. (1961), *British Towns: A Statistical Study of their Social and Economic Differences*. London: Oliver & Boyd Ltd.

Mosteller, F., et al. (1961), *Probability with Statistical Applications*. Reading, Mass.: Addison-Wesley Publishing Co., Inc.

Mosteller, F., and Wallace, D. L. (1964), *Inference and Disputed Authorship: The Federalist*. Reading, Mass.: Addison-Wesley Publishing Co., Inc.

Nagel, E. (1961), *The Structure of Science*. New York: Harcourt, Brace & World, Inc.

National Academy of Sciences—National Research Council (1965), *The Science of Geography*. Report of the Ad Hoc Committee on Geography, Earth Sciences Division.

Neft, D. S. (1967), "Statistical Analysis for Areal Distributions." *Monograph Series*, Regional Science Research Institute, No. 2.

Nelson, H. J. (1955), "A Service Classification of American Cities." *Economic Geography*, Vol. 31, pp. 189–210.

Neyman, J., Scott, E. L., and Shane, C. A. (1956), "Statistics of Images of Galaxies with Particular Reference to Clustering." *Proceedings of the Third Berkeley Symposium on Mathematical Statistics and Probability*, Vol. 3, pp. 75–111.

Neyman, J., and Scott, E. L. (1957), "On a Mathematical Theory of Populations Conceived as a Conglomeration of Clusters." *Cold Spring Harbor Symposia on Quantitative Biology*, Vol. 22, pp. 109–120.

Neyman, J. (1960), "Indeterminism in Science and New Demands on Statisticians." *Journal of American Statistical Association*, Vol. 55, pp. 625–639.

Nystuen, J. D. (1967), "A Theory and Simulation of Intraurban Travel." In W. L. Garrison and D. F. Marble (eds.), *Quantitative Geography*, Northwestern University Studies in Geography, No. 13, pp. 84–94.

O'Leary, M., Lippert, R. H., and Spitz, O. T. (1966), "Fortran IV and Map Program for Computation and Plotting of Trend Surfaces for Degrees 1 Through 6." *Computer Contribution 3*, State Geological Survey, University of Kansas.

Olsson, G. (1965), "Distance and Human Interaction. A Migration Study." *Geografiska Annaler*, Vol. 47, pp. 3–43.

Parzen, E. (1960), *Modern Probability Theory and Its Applications*. New York: John Wiley & Sons, Inc.

Parzen, E. (1962), *Stochastic Processes*. San Francisco: Holden-Day, Inc.

Pearson, E. S. (1963), "Comparison of Tests for Randomness of Points on a Line." *Biometrika*, Vol. 50, pp. 315–323.

Peixoto, J. P., Saltzman, B., and Teweles, S. (1964), "Harmonic Analysis of the Topography along Parallels of the Earth." *Journal Geophysical Research*, Vol. 69, pp. 1501–1505.

Peltier, L. C. (1962), "Area Sampling for Terrain Analysis." *The Professional Geographer*, Vol. 14, pp. 24–28.

Penrose, L. S. (1947), "Some Notes on Discrimination." *Ann. Eugenics*, Vol. 13, pp. 228–237.

Pielou, E. C. (1957), "The Effect of Quadrat Size on the Estimation of the Parameters of Neyman's and Thomas's Distributions." *Journal of Ecology*, Vol. 45, pp. 31–47.

Pielou, E. C. (1959), "The Use of Point-to-Plant Distances in the Study of the Pattern of Plant Populations." *Journal of Ecology*, Vol. 47, pp. 607–613.

Pielou, E. C. (1961), "Segregation and Symmetry in Two-Species Populations as Studied by Nearest-Neighbor Relations." *Journal of Ecology*, Vol. 49, pp. 255–269.

Pierson, W. F., et al. (1960), "The Directional Spectrum of a Wind Generated Sea as Determined from Data Obtained by the Stereo Wave Observation Project." *Meteorological Papers*, Vol. 2, New York University.

Pitts, F. R. (1963), "Problems in Computer Simulation of Diffusion." *Papers*, Regional Science Association, Vol. 11, pp. 111–119.

Plackett, R. L. (1966), "Current Trends in Statistical Inference." *Journal of the Royal Statistical Society*, Series A, Vol. 129, pp. 249–267.

Porter, P. W. (1963), "What is the Point of Minimum Aggregate Travel?" *Annals*, Association of American Geographers, Vol. 53, pp. 224–232.

Porter, P. W. (1964), "A Comment on 'The Elusive Point of Minimum Travel'." *Annals*, Association of American Geographers, Vol. 54, pp. 403–406.

Preston, F. W. (1966), "Two-Dimensional Power Spectra for Classification of Land Forms." In D. F. Merriam (ed.), "Computer Applications in the Earth Sciences: Colloquium on Classification Procedures." *Computer Contribution 7*, State Geological Survey, University of Kansas, pp. 64–69.

Price, D. O. (1942), "Factor Analysis in the Study of Metropolitan Centers." *Social Forces*, Vol. 20, pp. 449–455.

Proudfoot, M. J. (1942), "Sampling with Transverse Traverse Lines." *Journal of the American Statistical Association*, Vol. 37, pp. 265–270.

Quandt, R. E. (1964), "Statistical Discrimination Among Alternative Hypotheses and Some Economic Regularities." *Journal of Regional Science*, 5, pp. 1–23.

Quenouille, M. H. (1949), "Problems in Plane Sampling." *Annals of Mathematical Statistics*, Vol. 20, pp. 355–375.

Rao, C. R. (1948), "The Utilization of Multiple Measurements in Problems of Biological Classification." *Journal Royal Statistical Society*, B. Vol. 10, pp. 159–203.

Rao, C. R. (1952), *Advanced Statistical Methods in Biometric Research*. New York: John Wiley & Sons, Inc.

Rao, C. R. (1965), *Linear Statistical Inference and Its Applications*. New York: John Wiley & Sons, Inc.

Rayner, J. N. (1966), "Harmonic and Spectral Analysis and Their Applications in Geography." Unpublished manuscript, Department of Geography, Ohio State University.

Rayner, J. N. (1967), "Correlation Between Surfaces by Spectral Methods." In D. F. Merriam and N. C. Cocke (eds.), "Computer Applications in the Earth Sciences: Colloquium on Trend Analysis." *Computer Contribution 12*, State Geological Survey, University of Kansas, pp. 31–37.

Roberts, M. C., and Rumage, K. W. (1965), "The Spatial Variations in Urban Left-Wing Voting in England and Wales, 1951." *Annals*, Association of American Geographers, Vol. 55, pp. 161–178.

Robinson, A. H. (1956), "The Necessity of Weighting Values in Correlation of Areal Data." *Annals*, Association of American Geographers, Vol. 46, pp. 233–236.

Robinson, A. H., and Bryson, R. A. (1957), "A Method for Describing Quantitatively the Correspondence of Geographic Distributions." *Annals*, Association of American Geographers, Vol. 47, pp. 379–391.

Robinson, A. H. Lindberg, J. B., and Brinkman, L. W. (1961), "A Correlation and Regression Analysis Applied to Rural Farm Population Densities in the Great Plains." *Annals*, Association of American Geographers, Vol. 51, pp. 211–221.

Rodgers, A. (1957), "Some Aspects of Industrial Diversification in the United States." *Economic Geography*, Vol. 33, pp. 16–30.

Rogers, A. (1965), "A Stochastic Analysis of the Spatial Clustering of Retail Establishments." *Journal of American Statistical Association*, Vol. 60, pp. 1094–1103.

Rudin, W. (1964), *Principles of Mathematical Analysis*, Second edition. New York: McGraw-Hill Book Co., Chapter 1.

Rulon, P. J. (1951), "Distinctions Between Discriminant and Regression Analyses and a Geometric Interpretation of the Discriminant Function. *Harvard Educational Review*, Vol. 21, pp. 80–90.

Rummel, R. J. (1963), "Dimensions of Conflict Behavior Within and Between Nations." In A. R. Rapoport and L. von Bertalanffy (eds.), General Systems: *Yearbook for the Society for General Systems Research*, Vol. 8.

Rushton, G., Golledge, R. G., and Clark, W. A. V. (1967), "Formulation and Test of a Normative Model for the Spatial Allocation of Grocery Expenditures by a Dispersed Population." *Annals*, Association of American Geographers, Vol. 57, pp. 389–400.

Rodoman, B. B. (1967), "Mathematical Aspects of the Formalization of Regional Geographic Characteristics." *Soviet Geography*, Vol. 8, pp. 687–708.

Russett, B. M. (1968). "Delineating International Regions." in J. D. Singer (ed.), *Quantitative International Politics: Insights and Evidence*. New York: The Free Press, pp. 317–374.

Russwurm, L. H. (1964), "The Central Business District Retail Sales Mix, 1948–1958." *Annals*, Association of American Geographers, Vol. 54, pp. 524–536.

Sabbagh, M. E., and Bryson, R. A. (1962), "Aspects of the Precipitation Climatology of Canada Investigated by the Method of Harmonic Analysis." *Annals*, Association of American Geographers, Vol. 52, pp. 426–440.

Salisbury, N. E., and Hart, J. F. (1965), "Population Change in Middle Western Villages: A Statistical Approach." *Annals*, Association of American Geographers, Vol. 55, pp. 140–160.

Sampford, M. R. (1962), "Methods of Cluster Sampling With and Without Replacement for Clusters of Unequal Sizes." *Biometrika*, Vol. 49, pp. 27–40.

Saunders, D. R. (1950), *Practical Methods in the Direct Factor Analysis of Psychological Score Matrices*. Ann Arbor: University Microfilms, Inc.

Scheffé, H. (1959), *The Analysis of Variance*. New York: John Wiley & Sons, Inc.

Schlager, K. J. (1964), "Simulation Models in Urban and Regional Planning." *Technical Record*, Southeastern Wisconsin Regional Planning Commission.

Schnore, L. F. (1961), "The Statistical Measurement of Urbanization and Economic Development." *Land Economics*, Vol. 37, pp. 224–245.

Schuessler, K. F., and Driver, H. (1956), "A Factor Analysis of Primitive Societies." *American Sociological Review*, Vol. 21, pp. 493–499.

Seal, H. (1964), *Multivariate Statistical Analysis for Biologists*. New York: John Wiley & Sons, Inc.

Seymour, D. R. (1965), "IBM 7090 Program for Locating Bivariate Means and Bivariate Medians." *Technical Report No. 16*, Computer Applications in the Earth Sciences Project, Department of Geography, Northwestern University.

Shear, J. A. (1966), "A Set-Theoretic View of the Köppen Dry Climates." *Annals*, Association of American Geographers, Vol. 56, pp. 508–515.

Siegel, S. (1956), *Nonparametric Statistics: For the Behavioral Sciences*. New York: McGraw-Hill Book Co.

Simmons, J. (1964), "The Changing Pattern of Retail Location." *Research Paper 92*, Department of Geography, University of Chicago.

Simon, H. A. (1955), "On a Class of Skew Distribution Functions." *Biometrika*, Vol. 52, pp. 425–440.

Simon, H. A. (1957), *Models of Man*. New York: John Wiley & Sons, Inc.

Singer, H. W. (1936), "The 'Courbe des Populations:' A Parallel to Pareto's Law." *Economic Journal*, Vol. 46, pp. 254–263.

Skellam, J. G. (1952), "Studies in Statistical Ecology: I, Spatial Pattern." *Biometrika*, Vol. 39, pp. 346–362.

Skokal, R. R., and Sneath, P. H. (1963), *Principles of Numerical Taxonomy*. San Francisco: W. H. Freeman and Co.

Smith, R. H. T. (1965), "Method and Purpose in Functional Town Classification." *Annals*, Association of American Geographers, Vol. 55, pp. 539–548.

Snedecor, G. W. (1956), *Statistical Methods*. Ames, Iowa: The Iowa State College Press.

Speight, J. G. (1965), "Meander Spectra of the Angabunga River." *Journal of Hydrology*, Vol. 3, pp. 1–15.

Stafford, H. A. (1960), "Factors in the Location of the Paperboard Container Industry." *Economic Geography*, Vol. 36, pp. 260–266.

Steger, W., and Douglas, N. J. (1964), "Simulation Model." *Progress Report No. 5*, Community Renewal Program, Department of City Planning, Pittsburgh.

Steindl, J. (1965), *Random Processes and the Growth of Firms*. New York: Hafner Publishing Company.

Steiner, D. (1965), "Die Faktorenanalyse; ein modernes statistisches Hilfsmittel des Geographen für die objektive Raumgliederung und Typenbildung." *Geographica Helvetica*, Vol. 20, pp. 20–34.

Sternstein, L. (1962), "Note on the Rank Correlation Method." *The Professional Geographer*, Vol. 14, pp. 10–12.

Stevens, S. S. (1946), "On the Theory of Scales of Measurement." *Science*, Vol. 103, pp. 677–680.

Stone, R. (1960), "A Comparison of the Economic Structure of Regions Based on the Concept of Distance." *Journal of Regional Science*, Vol. 2, pp. 1–20.

Stone, R. (1966), *Mathematics in the Social Sciences and Other Essays*. Cambridge, Mass.: M.I.T. Press.

Stouffer, S., et al. (1950), *Measurement and Prediction*. Princeton: Princeton University Press.

Strahler, A. N. (1954), "Statistical Analysis in Geomorphic Research." *Journal of Geology*, Vol. 62, pp. 1–25.

Suits, D. B. (1957), "Use of Dummy Variables in Regression Equations." *Journal of the American Statistical Association*, Vol. 52, pp. 548–551.

Sumner, A. R. (1953), "Standard Deviation of Mean Monthly Temperatures in Anglo-America." *Geographical Review*, Vol. 43, pp. 50–59.

Taaffe, E. J. (1958), "A Map Analysis of United States Airline Competition. Pt. II—Competition and Growth." *The Journal of Air Law and Commerce*, Vol. 25, pp. 402–427.

Taaffe, E. J. (1959), "Trends in Airline Passenger Traffic: A Geographic Case Study." *Annals*, Association of American Geographers, Vol. 49, pp. 393–408.

Taaffe, E. J., Morrill, R. L. ,and Gould, P. R. (1963), "Transport Expansion in Underdeveloped Countries: A Comparative Analysis." *Geographical Review*, Vol. 53, pp. 503–529.

Taaffe, E. J., Garner, B. J., and Yeates, M. H. (1963), *The Peripheral Journey to Work: A Geographic Consideration*. Evanston: The Northwestern University Press.

Tatsuoka, M. M., and Tiedeman, D. V. (1954), "Discriminant Analysis." *Review Educational Research*, Vol. 24, pp. 402–420.

Technometrics. (1961), Vol. 3. Special issue devoted to spectral analysis.

Thomas, E. N. (1960a), "Areal Associations Between Population Growth and Selected Factors in the Chicago Urbanized Area." *Economic Geography*, Vol. 36, pp. 158–170.

Thomas, E. N. (1960b), "Maps of Residuals from Regressions: Their Characteristics and Uses in Geographic Research." *Monograph No. 2*, Department of Geography, University of Iowa.

Thomas, E. N. (1961), "Toward an Expanded Central Place Model." *Geographical Review*, Vol. 51, pp. 400–411.

Thomas, E. N. (1962), "The Stability of Distance-Population Size Relationships for Iowa Towns from 1900 to 1950." *Lund Studies in Geography, Series B*, Vol. 24, pp. 13–30.

Thomas, E. N. (1967), "Additional Comments on Population Size Relationships for Sets of Cities." In W. L. Garrison and D. F. Marble (eds.) *Quantitative Geography*, Northwestern University Studies in Geography, No. 13, pp. 167–190.

Thomas, E. N., and Anderson, D. L. (1965), "Additional Comments on Weighting Values in Correlation Analysis of Areal Data." *Annals*, Association of American Geographers. Vol. 55, pp. 492–505.

Thompson, H. R. (1955), "Spatial Point Processes with Applications to Ecology." *Biometrika*, Vol. 42. pp. 102–115.

Thompson, H. R. (1956), "Distribution of Distance to Nth Neighbor in a Population of Randomly Distributed Individuals." *Ecology*, Vol. 37, pp. 391–394.

Thompson, J. H. (1955), "A New Method for Measuring Manufacturing." *Annals*, Association of American Geographers, Vol. 45, pp. 416–436.

Thompson, J. H., Sufrin, S. C., Gould, P. R., and Buck, M. A. (1962), "Toward a Geography of Economic Health: The Case of New York State." *Annals*, Association of American Geographers, Vol. 52, pp. 1–20.

Tobler, W. R. (1964), "A Polynomial Representation of Michigan Population." *Papers*, Michigan Academy of Science, Arts, and Letters, Vol. 49, pp. 445–452.

Tobler, W. R. (1965), "Computation of the Correspondence of Geographical Patterns." *Papers*, Regional Science Association, Vol. 14, pp. 131–139.

Tobler, W. R. (1966a), "Numerical Map Generalization and Notes on the Analysis of Geographical Distributions." *Dicsussion Paper No. 8*, Michigan Inter-University Community of Mathematical Geographers.

Tobler, W. R. (1966b), "Spectral Analysis of Spatial Series." *Proceedings*, Fourth Annual Conference on Urban Planning Information Systems and Programs, Berkeley, pp. 179–185.

Tocher, K. D. (1963), *The Art of Simulation*. London: The English Universities Press, Ltd.

Tolstov, G. P. (1962), *Fourier Series*. Englewood Cliffs N.J.: Prentice-Hall, Inc.

Torgerson, W. S. (1958), *Theory and Methods of Scaling*. New York: John Wiley & Sons, Inc.

Torgerson, W. S. (1965), "Multidimensional Scaling of Similarity." *Psychometrika*, Vol. 30, pp. 379–393.

Törnqvist, G. (1967), *TV-agandets utveckling i Sverige 1956-65*. Stockholm: Almqvist and Wiksell.

Tukey, J. W. (1949), *The Sampling Theory of Power Spectrum Estimates*. Symposium on Applications of Autocorrelation Analysis to Physical Problems, Woods Hole, Mass.

Ullman, E. L., and Dacey, M. F. (1962), "The Minimum Requirements Approach to the Urban Economic Base." *Lund Studies in Geography, Series B*, Vol. 24, pp. 121–143.

United States Department of Agriculture (1950), "Generalized Types of Farming in the United States." *Agriculture Information Bulletin*, No. 3.

Veitch, L. G. (1965), "The Description of Australian Pressure Fields by Principal Components." *Quarterly Journal*, Royal Meteorological Society, Vol. 91, pp. 184–195.

Ward, J. H. (1961), "Hierarchical Grouping to Maximize Payoff." *Technical Report WADD-TN-61-29*, U.S.A.F., Lackland Air Force Base, Texas.

Ward, J. H. (1963), "Hierarchical Grouping to Optimize an Objective Function." *Journal American Statistical Association*, Vol. 58, pp. 236–244.

Warntz, W., and Stewart, J. Q. (1959), "Some Parameters of the Geographical Distribution of Population." *Geographical Review*, Vol. 49, pp. 270–272.

Warntz, W., and Neft, D. (1960), "Contributions to a Statistical Methodology for Areal Distributions." *Journal of Regional Science*, Vol. 2, pp. 47–66.

Weaver, J. C. (1954), "Crop-Combination Regions in the Middle West." *Geographical Review*, Vol. 44, pp. 175–200.

Whitten, E. H. T. (1963), "A Surface-Fitting Program Suitable for Testing Geological Models Which Involve Areally-Distributed Data." *Technical Report No. 2*, Computer Applications in the Earth Sciences Project, Department of Geography, Northwestern University.

Wilks, S.S. (1948), "Order Statistics." *Bulletin of the American Mathematical Society*, Vol. 54, pp. 6–50.

Williams, R. M. (1956), "The Variance of the Mean of Systematic Samples." *Biometrika*, Vol. 43, pp. 137–148.

Williamson, E., and Bretherton, M.H. (1963), *Tables of the Negative Binomial Probability Distribution*. New York: John Wiley & Sons, Inc.

Wold, H., and Jureen, L. (1953), *Demand Analysis*. New York: John Wiley & Sons.

Wolpert, J. (1965), "Behavioral Aspects of the Decision to Migrate." *Papers*, Regional Science Association, Vol. 15, pp. 159–169.

Wong, S. T. (1963), "A Multi-variate Statistical Model for Predicting Mean Annual Flood in New England." *Annals*, Association of American Geographers, Vol. 53, pp. 298–311,

Wood, W. F. (1955), "Use of Stratified Random Samples in a Land Use Study." *Annals*, Association of American Geographers, Vol. 45, pp. 350–367.

Wright, J. K. (1937), "Some Measures of Distributions." "*Annals*, Association of American Geographers, Vol. 27, pp. 177–211.

Yates, F. (1953), *Sampling Methods for Censuses and Surveys*. London: Charles Griffin & Co., Ltd.

Yeates, M. H. (1965), "Some Factors Affecting the Spatial Distribution of Chicago Land Values, 1910–1960." *Economic Geography*, Vol. 41, pp. 57–70.

Yuill, R. S. (1964), "A Simulation Study of Barrier Effects in Spatial Diffusion Studies." *Technical Report No. 1*, Spatial Diffusion Study, Department of Geography, Northwestern University.

Yule, G. U., and Kendall, M. G. (1950), *An Introduction to the Theory of Statistics.* New York: Hafner Publishing Co.

Zobler, L. (1957), "Statistical Testing of Regional Boundaries." *Annals*. Association of American Geographers, Vol. 47, pp. 83–95.

Zobler, L. (1958), "Decision Making in Regional Construction." *Annals*, Association of American Geographers, Vol. 48, pp. 140–148.

Zobler, L. (1959), "The Distinction Between Relative and Absolute Frequencies in Using Chi-Square for Regional Analysis." *Annals*, Association of American Geographers. Vol. 49, pp. 456–457.

INDEX